Third Generation Photovoltaic Technology
Recent Progress and Future Perspectives

Edited by

Alagarsamy Pandikumar[1], G. Murugadoss[2]

[1]Materials Electrochemistry Division, CSIR-Central Electrochemical Research Institute, Karaikudi-630003, Tamil Nadu, India

[2]Centre for Nanoscience and Nanotechology, Sathyabama Institute of Science and Technology, (Deemed to be University), Rajiv Gandhi Salai, Chennai - 600 119. Tamilnadu, India

Published by **Materials Research Forum LLC**
Millersville, PA 17551, USA

Published as part of the book series
Materials Research Foundations
Volume 163 (2024)
ISSN 2471-8890 (Print)
ISSN 2471-8904 (Online)

Print ISBN 978-1-64490-302-5
eBook ISBN 978-1-64490-303-2

Distributed worldwide by

Materials Research Forum LLC
105 Springdale Lane
Millersville, PA 17551
USA
https://www.mrforum.com

Manufactured in the United States of America
10 9 8 7 6 5 4 3 2 1

Table of Contents

Preface

Solution-processed third-generation solar cells (SCs) are built on inorganic nanoparticles, hybrids, or semiconducting organic macromolecules. This book focuses on various forms of third-generation solar cells, including Dye-sensitized solar cells, Polymer/Organic solar cells, Copper Zinc Tin Sulfide thin film solar cells, Quantum Dots solar cells and Perovskite-based cells.

The fundamentals, mechanism, and key information of third-generation solar cells concerning efficiency enhancement are thoroughly covered in this edited book, "Third Generation Photovoltaic Technology: Recent Progress and Future Perspectives".

Additionally, it covers the device architecture, interface engineering, stability concerns with third-generation solar cells, and the fabrication process, which includes preparation of the electrode film, absorber layer and the hole transport layer. The significance of flexible substrates like polyethylene terephthalate (PET) and polymide (PI) as well as semiconductor electrode materials like TiO_2, SnO_2, ZnO, and NiO is covered as well.

The introduction of each chapter provides a brief assessment of the different types of third-generation solar cell as well as the current approach used for each type of solar cell. Along with covering all the latest advances in energy conversion, electrode materials, and effective light-absorbing materials. The book further focuses on advanced solar cell applications at present and for future prospective.

The chapters cover different types of solar cells, how they are fabricated, and how the device's contact layer, absorber layer, and charge collection electrodes align with respect to their respective bang gaps.

In order to achieve sustainable practices, the book covers the most crucial characterization and fabrication techniques for all third-generation solar cells, including spin coating, blade coating, slot-die coating, dip coating, meniscus coating, spray coating, ink-jet printing, screen printing, electro deposition, and more.

Discussions are held regarding improvements in various components and performance metrics such fill factor, open circuit voltage, conversion efficiency, and short-circuit current density.

For academics, engineers, and business researchers seeking the most recent advancements in the industry, this book is an invaluable resource. In order for early career researchers to succeed in the upcoming decade, we believe that the book will help them build their own research direction in the field of third generation energy conversion.

Third Generation Photovoltaic Technology
Materials Research Foundations 163 (2024) 1-37

Materials Research Forum LLC
https://doi.org/10.21741/9781644903032-1

Chapter 1

The Future of Solar Energy: A Deep Dive into Third Generation Solar Cells

A. Murugeswari[1], I.V. Asharani[2], N. Arunai Nambi Raj[3]

[1]Department of Physics, Anna University, Chennai -600 025

[2]Department of Chemistry, School of Advanced Sciences, Vellore Institute of Technology - 632 014, India

[3]Department of Physics, School of Advanced Sciences, Vellore Institute of Technology – 632 014, India

Abstract

This chapter provides a comprehensive discussion of the developments in solar cells encompassing first, second, and third-generation solar cells. Specifically, the focus lies on third-generation solar cells, including an overview of their progression and a performance comparison with other genres of solar cells. The operational principles and materials used in photovoltaic studies are elaborated. Moreover, the chapter explores recent breakthroughs aimed at enhancing the efficiency of DSSC components, organometal halide perovskite compounds, and organic solar cells.

Keywords

Solar Cells, Third Generation, DSSC, Charge Transfer, Perovskite Solar Cells, Organic Solar Cells, Electron Transport Layer

Contents

Materials Research Forum LLC
https://doi.org/10.21741/9781644903032-1

1. Introduction

Global energy demand is significantly increasing every year, prompting researchers worldwide to explore clean methods of energy production to keep up with this rising consumption. Among both renewable and non-renewable energy sources, the sun emerges as the foremost energy source and it is highly utilized across the globe, with many countries offering incentives to setup solar energy based power sources. For the purpose of utilizing solar energy to meet human requirements, there are diverse thermodynamic routes available allowing the conversion of solar energy into heat, kinetic, electrical and chemical energy. Photovoltaic (PV) technology involves the direct conversion of radiation into electricity. Utilizing photovoltaic cells, commonly known as solar cells, harnesses sunlight and converts it into electrical energy through the photoelectric effect and they are highly regarded as promising devices for energy generation. Sunlight contains solar energy radiation in the form of photons, each carrying varying frequency of energy within the solar spectrum. Upon reaching a solar cell, sunlight encounters a phenomenon where certain photons are absorbed while others are reflected. The absorbed photons impart enough energy to dislodge electrons from their atoms within the solar cell material. Subsequently, these liberated electrons migrate towards the top surface of the solar cell, specially devised to enhance their receptivity to free electrons. As multiple electrons with a negative charge accumulate at the cell's front surface, an imbalance of charge arises between the top and rear surfaces, resulting in the development of an electric potential [1].

Alexandre-Edmund Becquerel is recognized as the first person to observe "electrical current arising from certain light induced chemical reactions" in 1839 [5]. Similar effects were later observed by other scientists using a solid material (selenium) several decades afterward. However, it was not until 1882 that the first practical PV cell was successfully developed by Charles Frittis [2]. This solar cell utilized gold thin film-coated selenium sheets [3]. A significant milestone occurred in 1954 when the Bell Laboratory constructed the first crystal PV cell with an efficiency of 4%. This indicated that merely 4% of the solar energy was successfully transformed into electrical energy [4].

The Power Conversion Efficiency (PCE) is used to evaluate a solar cell's effectiveness in converting sunlight into electrical power. PCE is defined as the ratio of the energy output from the solar cell to the input energy received from the sun (P_{in}) and can be calculated using the equation: PCE = (V_{oc} x J_{sc} x FF / P_{in}) x 100. It is determined by three essential parameters: the open-circuit voltage (V_{oc}), short-circuit current (J_{sc}), and fill factor (FF). A higher PCE signifies a more efficient conversion of sunlight into electrical energy. V_{oc} represents the potential difference across a solar cell when it is exposed to light without any external load and it is primarily determined by the bandgap of the semiconductor material. FF is the ratio of the maximum power generated by the solar cell (P_{max}) to the product of V_{oc} and J_{sc}, which can be calculated using the equation: FF = P_{max} / (V_{oc} x J_{sc}). Graphically, FF measures the "squareness" of the solar cell's current density-voltage (J-V) curve, indicating how close the curve resembles a rectangle under light exposure. Scientists continuously explore various methods to optimize these parameters and improve solar cell efficiency in real-world conditions. Based on scientific investigations on solar cells, these devices are categorized into three generations. The first generation of PV cells revolves around single-junction crystal solar cells based on silicon wafers, including single and polycrystalline silicon. Moving on to the second generation, the focus shifts to single-junction devices like CdTe, CIGS, and a-Si, with the aim of optimizing material usage while maintaining earlier achieved efficiencies. This generation strives to reduce material costs by utilizing thinner films. In contrast, the third generation takes a different approach, focusing on double, triple, multi-junction, and nanotechnology based solar cells. These advancements show promising results for achieving efficient cells at a lower cost. Hence, researchers worldwide are motivated to explore novel materials and innovative techniques to create solar devices that are not only efficient but also sustainable and capable of large-scale production.

2.   Types of solar cells

2.1   First generation

The first generation solar cells are based on silicon (Si) wafers, including Si-single crystals and bulk polycrystalline Si wafers, due to Si's suitability as a semiconductor with an energy gap of 1.1 eV for photovoltaic applications. Si-single crystal solar cells have an efficiency range of 14–19%. These crystalline silicon solar cells are classified into three types: mono-crystalline silicon (mono-c-Si), poly-crystalline silicon (poly-c-Si), and ribbon silicon [5]. Among them, mono-c-Si exhibits the highest efficiency, reaching up to 27%, utilizing high-purity crystalline silicon structures obtained from silicon rods sliced into thin wafers for PV solar cell fabrication with an expected lifespan of 25 to 30 years. Conversely, poly-c-Si-based PV cells, consisting of numerous small grains, provide more efficient production compared to mono-c-Si PV cells but generally have lower efficiencies due to border defects on the crystal structures. Both single crystal and polycrystalline silicon are widely used as they are abundant and non-toxic [6]. However, to meet increasing demand and reduce

Materials Research Forum LLC
https://doi.org/10.21741/9781644903032-1

costs, polycrystalline silicon is favoured for commercial modules, despite its slightly lower efficiency. The higher cost and delicate technical procedure involved have led to the use of polycrystalline Si instead of single crystal wafers, albeit at the expense of a slightly reduced solar conversion efficiency. Presently, silicon (Si) based solar cells have achieved a PCE of 26% with a lifespan exceeding 20 years.

2.2 Second generation

Extensive research has driven the advancement of second generation solar cells, which are built on thin film technology. These solar cells consist of multiple thin layers, typically ranging from 1 to 4 μm in thickness, carefully deposited onto diverse substrates such as polymer, glass, cadmium, or metallic materials [7]. Thin film based solar cells can be broadly categorized into four types: amorphous silicon, cadmium telluride (CdTe), copper indium selenide, and copper indium gallium diselenide (CIGS). Among these technologies, CdTe–based solar cells have grown as the most significant commercial thin-film modules globally. Due to the high production costs associated with crystalline silicon, thin film cells based on a hydrogenated alloy of amorphous silicon have been successfully commercialized. These cells are now available on the market and exhibit solar conversion efficiencies ranging from 12% to 16%, influenced by specific manufacturing procedures and wafer quality. The multi-junction concept allows for the incorporation of various materials with different bandgaps. Notably, CIGS solar cells offer the highest efficiencies among thin film–based solar cells, with current efficiencies varying from 7% to 16%. In laboratory tests, CIGS cells have achieved efficiencies up to 20%, placing them in close competition with crystalline silicon–based solar cells. The advantages of thin film technology include a low static load characteristic due to the lightweight nature of the cells and reduced manufacturing costs compared to silicon wafer based cells, which require more materials. Thin film modules can be produced in large sizes, further adding to their appeal. However, it is essential to consider that thin film solar cells generally exhibit lower efficiencies compared to wafer based cells. Additionally, the use of certain materials, such as cadmium, raises environmental concerns due to their toxicity, and the availability of tellurium is relatively limited compared to cadmium.

2.3 Third generation

Third generation solar cells are emerging as viable alternatives to both Si-based solar cells and thin film solar cells. Despite their current limitations in conversion efficiency and photovoltaic performance stability, they show promise for specific applications. Notably, their ability to be fabricated as thin, lightweight, and flexible solar modules makes them ideal for portable electronics. Additionally, their high efficiency under dim light conditions outperforms the existing technologies, making them well-suited for ambient energy harvesting in wireless sensors used for the internet of things (IoT) devices. Perovskite solar cells (PSCs), organic solar cells (OSCs), and dye-sensitized solar cells (DSSCs) represent examples of third-generation solar cells. These technologies are currently at the pre-commercial stage, ranging from the demonstration (multi-junction) stage to novel concepts in the quantum structured PV stage. Of particular interest, these third-generation PV solar cells have the potential to surpass the Shockley-Queisser (SQ) limit, which states that conventional solar cells can only achieve a maximum conversion efficiency of 31% [8]. In addition to the emerging third generation solar cells such as DSSCs, PSCs, OSCs, combination of a range of PV solar cells made up of silicon (first-generation) and thin film (second-generation) also materialized, which includes quantum dot solar cell (QDSC), copper zinc tin sulfide solar cell (CZTS), and tandem/multijunction cell (MJC) to overcome the SQ limit.

2.4 CZTS solar cells

CZTS solar cells rely on two synthetic compounds: copper zinc tin sulphide (Cu_2ZnSnS_4) (CZTS) and copper zinc tin selenide ($Cu_2ZnSnSe_4$) (CZTSe). The optical and electronic properties of CZTS and CZTSe are comparable to those of CdTe and CIGS, making CZTS solar cells an intriguing alternative to CIGS and CdTe solar cell. CZTS components, namely copper, tin, zinc, and sulfur, are environmentally friendly, nontoxic and abundant. These solar cells are fabricated using thin film solar technology, utilizing a quaternary semiconducting compound [9].

One of the significant advantages of CZTS is considerably lower raw material cost compared to the three currently existing thin film PV technologies: c-Si, CdTe, and CIGS. CZTS solar cells can serve as absorbers and are divided into two structures: stannite type and kesterite type. The latter, kesterite type, offers reduced toxicity and a higher number of components. The most efficient kesterite solar cell achieves an efficiency of 10.1%, and researchers have enhanced its efficiency through various strategies, such as optimizing the deposition process and refining the interface between the kesterite absorber and the buffer layer [10].

2.5 Quantum dots

Quantum dots (QDs) represent a unique class of semiconductor nanoparticles with dimensions below 10 nm, comprising periodic groups of II-VI, III-V, or IV-VI materials. Their properties can be easily manipulated by altering their size and shape. QD solar cells are structured with a tunable bandgap, enabling them to match the spectral distribution of the solar spectrum, thereby reducing the cost-to-watt ratio of solar electricity. These cells efficiently capture a broad range of the solar spectrum [11].

QDs offer several advantages, including their adaptability to be moulded into various forms, either in two-dimensional sheets or three-dimensional arrays. They can be processed to form junctions on cost-effective substrates such as plastics, glass, or metal sheets. Moreover, QDs can seamlessly integrate with organic polymers and dyes. Achieving a PCE of 16%, these cells hold the potential for commercialization as flexible and portable PVs. To enhance energy conversion performance, researchers have employed CdS QDs, CdSe, and CdSe/CdS core-shell QDs.

2.6 Multijunction solar cells

Amorphous silicon and compound semiconductors from group III and V elements of the periodic table have demonstrated significant improvements in efficiency. However, their manufacturing processes remain costly. To address this, gallium arsenide (GaAs), a compound semiconductor, has emerged as a compelling alternative to silicon due to its high efficiency. GaAs solar cells are available in various forms, including thin film, single crystal, multicrystalline, and multijunction. In laboratory settings, thin film GaAs single junction solar cells have exhibited an impressive efficiency of 28.8%, while multicrystalline GaAs solar cells have shown an efficiency of 18.4%. Additionally, thin film GaAs solar cell modules have reached 24.1% efficiency in laboratory conditions [12]. The use of multijunction solar cells incorporating GaAs has achieved even higher efficiencies, reaching 31.6%. Beyond its exceptional efficiency, GaAs possesses other advantageous characteristics. i.e. bandgap of 1.42eV, exhibits high resilience in high-temperature environments and superior resistance to radiation. These properties make GaAs well suited for use in space applications and solar concentrators.

3. Classification of third generation solar cells

Three different types of third generation solar cells emerged in the last two decade includes,

(i) Dye-sensitized solar cells,

(ii) Perovskite solar cells,

(iii) Organic solar cells.

The capture and conversion of light energy in these solar cells is made possible through the modification of a nanostructured semiconductor interface using dyes, conjugate polymers, or semiconductor nanocrystals. However, enhancing the efficiency of photoinduced charge separation and the transportation of charge carriers across these

Third Generation Photovoltaic Technology   Materials Research Forum LLC
Materials Research Foundations 163 (2024) 1-37   https://doi.org/10.21741/9781644903032-1

nanoassemblies remains a significant challenge. These solar cells are discussed in the following sections.

3.1   Dye-sensitized solar cells (DSSC)

In 1991, the discovery of DSSCs was credited to O'Regan and Grätzel [13]. These solar cells have garnered commercial interest due to their semi-transparency, cost-effectiveness, ease of fabrication, and satisfactory performance in lowlight conditions. DSSCs, also known as Grätzel cells, are photochemical solar cells that use an electrolyte as a medium to convert sunlight into electricity [14-16]. The cell is composed of a photoanode, sensitizer, electrolyte, and counter electrode. Over the last two decades, significant advancements have led to the development of various novel DSSC devices, stemming from the original Grätzel cell prototype. These variants include P-type, tandem, hybrid, wire form, solid-state, and quasi-solid-state DSSCs. For instance, the dye can be replaced with quantum dots to create a Quantum Dot Sensitized Solar Cell (QDSSC). Similarly, the liquid electrolyte can be gelated (quasi-solid-state DSSC) or replaced by a solid hole conductor (solid-state DSSC). Among these possibilities, the tandem DSSC stands out for its potential to achieve high power conversion efficiency. Recent developments in DSSCs have spurred numerous research endeavors aiming to achieve higher efficiency.

3.1.1   DSSC Unit

The DSSC structure comprises several components such as conductive substrates, photoanode/photoelectrode, dye/sensitizer, electrolyte, and catalyst/counter electrode etc. The basic structure and component of DSSC is shown in Fig 1.

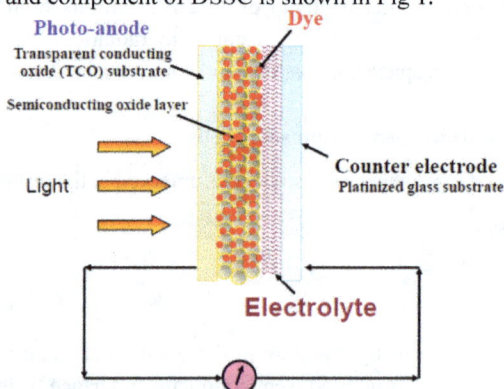

Figure 1: Schematic diagram of a DSSC and its components adapted from ref (20)

3.1.2 Substrate

The conventional fabrication of DSSCs involves utilizing two transparent conducting oxide (TCO) coated glass substrates. These substrates serve as a foundation for depositing the semiconductor and catalyst materials. In order to maximize sunlight absorption on the active area of the cell, the substrate must possess both high transparency and excellent conductivity [17]. For optimal light absorption, the transparency of the substrate should exceed 80%. Typically, glass is chosen as the conductive substrate, with two common options being Fluorine Tin Oxide (FTO) and Indium Tin Oxide (ITO). Both FTO and ITO consist of soda lime glass coated with layers of indium and fluorine tin oxide, respectively [18]. To meet the specific requirements, different substrate materials are employed in DSSC fabrication, focusing on two crucial properties: conductivity and transparency.

3.1.3 Photo electrode

Semiconductor metal oxides play a crucial role as the photoelectrode or photoanode in DSSCs by acting as electron collectors. The overall PCE of DSSCs is heavily influenced by the morphology and band-gap of the photoanode material. This material's primary function is to adsorb dye and efficiently transfer electrons to the external circuit [19].

For a photoanode material to be efficient, it should possess the following key characteristics: i) High photo corrosion resistance to ensure long term stability. ii) A large surface area to maximize dye absorption. iii) Excellent light diffusion, a rough surface, and a high capability to accept electrons. iv) The bandgap of the photoanode material directly impacts the overall efficiency, so optimization is essential. v) Strong adhesion with the substrates to maintain structural integrity [21].

3.1.4 Semiconductor materials

Figure 2 shows several popular metal oxide examples, including TiO_2, ZnO, SnO_2, $SrTiO_3$, Zn_2SnO_4, WO_3, and Nb_2O_5, all of which possess higher band gaps (suffix in g > 3eV). These semiconductor materials are commonly utilized as photoanode materials due to their superior photo corrosion resistance and remarkable electronic properties.

To enhance the efficiency of mesoporous photoanode thin films, researchers are exploring the incorporation of carbon allotropes and conducting polymers. These materials offer excellent light-scattering properties and promote good electrical conductivity, making them ideal candidates to produce highly efficient photoanodes for DSSCs.

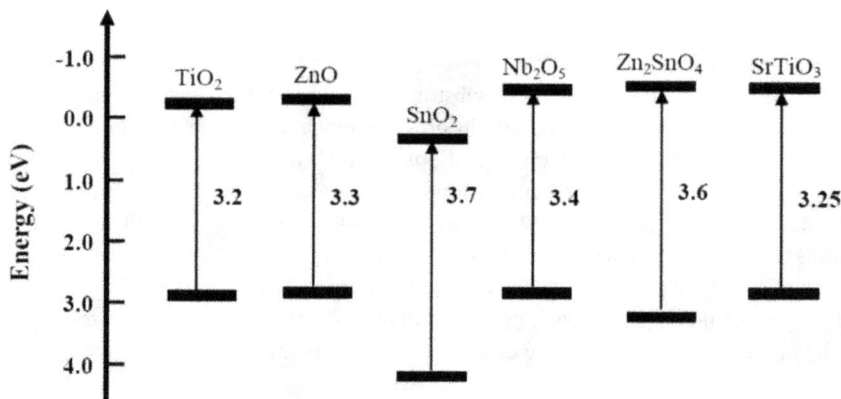

Figure 2: Band positions of the most common semiconductors adapted from ref [20]

3.1.6 Dye (photosensitizer)

The photosensitizer plays a vital role in DSSCs, contributing two essential characteristics: direct sunlight absorption and the transfer of excited electrons to the semiconductor's conduction band. Its significance lies in generating photogenerated electrons and injecting them into the TiO_2 semiconductor's conduction band, benefiting both p-type and n-type DSSCs.

An optimal photosensitizer must satisfy the following prerequisites:

i. Its absorption spectrum should span a broad range within the visible region. ii. The dye should possess robust anchoring groups to enhance its adhesion to the semiconductor surface. iii. The dye's lowest unoccupied molecular orbital (LUMO) should exceed the semiconductor's conduction band, facilitating swift charge transfer. iv. The dye's regeneration must be sustained, with its oxidized state level surpassing the redox potential of the electrolyte. v. The photosensitizer's stability should permit around 108 turnovers, equivalent to two decades of efficient DSSC operation. vi. The dye should demonstrate stability in terms of electrochemistry, thermal conditions, and exposure to light.

Dye sensitizers can be categorized into two types: 1) metal complex based dyes and 2) organic dyes. Among metal complexes, ruthenium based metal complexes, particularly N3, N719, and N749, are preferred due to their exceptional photovoltaic properties. These Ru-complexes offer a wide range of absorption spectra, suitable energy level states, and relatively long excited state lifetimes, leading to a maximum light conversion efficiency of 13.0% and holding great potential for advancing DSSC technology.

Additionally, metal free organic sensitizers have also been utilized in DSSCs, achieving an impressive PCE of 14.3%. These organic dyes can be further classified based on their spectral response as visible light responsive and Near-Infrared (NIR) light responsive, including porphyrins and squaraine [22-25].

3.1.7 Electrolyte

The overall durability of DSSCs hinges on the careful selection of the electrolyte, which plays a pivotal role in rejuvenating the dye following the injection of electrons into the semiconductor's conduction band and the transfer of holes towards the counter electrode [26]. By utilizing a reduced state redox species, the electrolyte regenerates oxidized dyes at the interface between the photoanode and the electrolyte. Furthermore, it facilitates the flow of electricity between the photoanode and the counter electrode through its ions and redox species, undergoing regeneration at the counter electrode with the assistance of electrocatalysts.

For DSSCs to attain efficiency and steadfast stability, the electrolyte must exhibit the following attributes [27]: i) A markedly positive redox potential to yield elevated voltage and bolster electrical conductivity. ii) Low viscosity to facilitate rapid electron transfer. iii) Strong interfacial affinity with the semiconductor, counter electrode, and nanocrystalline materials. iv) Minimal absorption of visible spectrum sunlight.

Liquid electrolytes, categorized as organic and ionic electrolytes based on their solvents, find common usage in DSSCs [28]. The iodide/triiodide (I^-/I^{3-}) redox couple, when combined with salt and a volatile solvent, is frequently employed due to its favorable redox potential that maximizes cell efficiency. Nevertheless, DSSCs relying on liquid electrolytes often grapple with limited stability owing to their high volatility and susceptibility to leakage.

To address these concerns regarding stability, solid state electrolytes and polymer gel electrolytes (PEGs) have been developed. PEGs typically comprise a foundational polymer, salt, and iodine. Further classification reveals thermoplastic polymer gel electrolytes (TPEGs) and thermoset polymer gel electrolytes (TSPGs). TPEGs feature physically interconnected polymer chains that form a network, while the presence of polymer content in both TPEGs and TSPGs enhances stability by curbing vaporization and liquid seepage. This advancement leads to improved performance and durability of DSSCs.

3.1.8 Counter electrode

The electrolyte within DSSCs assumes a pivotal role in orchestrating the transfer of electrons to dye molecules, consequently triggering their oxidation. Following this process, the oxidized electrolyte undergoes diffusion toward the counter electrode (CE), where reduction reactions take place. The CE boasts a dual purpose function: absorbing and conveying electrical energy into and out of the cell. To effectively fulfill this task, the CE must exhibit minimal resistance while also operating as a catalyst for the reduction of the oxidized form of the redox mediator.

The preparation of the CE involves the application of a catalytic layer composed of platinum onto a conducting glass substrate. Since glass inherently possesses poor electrode characteristics, it introduces substantial charge transfer resistance, particularly within standard iodide/triiodide/electrolyte systems [29]. Platinum (Pt) catalyst stands as the favored option for the CE due to its elevated exchange current density, exceptional catalytic process, and transparency. Nonetheless, researchers have been delving into alternate materials to Pt, encompassing non-Pt metals, transition metal compounds, conducting polymers, and carbonaceous substances [30]. These prospective alternatives aim to either uphold or enhance the catalytic attributes of the CE while potentially yielding cost savings and enhanced efficiency for DSSCs.

3.1.9 Operation mechanism

The DSSC structure (Fig.3) typically comprises two FTO-coated glass substrates. One substrate is covered with a nanocrystalline semiconductor layer (usually TiO₂ particles) functioning as the photo-electrode (PE) when sensitized with a dye, commonly a ruthenium (Ru) based organometallic molecule. The other substrate, coated with a catalyst like Pt, serves as the CE. The PE and CE are either laminated together with a thermoplastic spacer foil or separated by a thick and porous insulator to avoid short circuit.

The DSSC must meet specific requirements to achieve efficient solar energy conversion:

1. To facilitate the seamless transfer of charges between the dye and the semiconductor, it is imperative that the energy levels of the dye's LUMO and highest occupied molecular orbital (HOMO) are situated at a more negative (higher) position than the CB and VB levels of the semiconductor, respectively.

2. For the uninterrupted progression of electron transfer, the CB level of the semiconductor should exhibit a more negative value in comparison to the conductive glass (either FTO or ITO).

3. Once the electron reaches the conductive glass, its departure into the external circuit and subsequent journey to the counter electrode are essential steps, finalizing the conversion of solar energy into electrical energy [31].

4. In order to sustain numerous cycles of energy conversion within the DSSC, both the electrolyte and the dye must undergo regeneration during each energy cycle. The counter electrode assumes the role of receiving electrons and rejuvenating the electrolyte; thus, it necessitates a LUMO level more negative than that of the dye (in accordance with NHE) to proficiently accommodate electrons and facilitate the dye's regeneration.

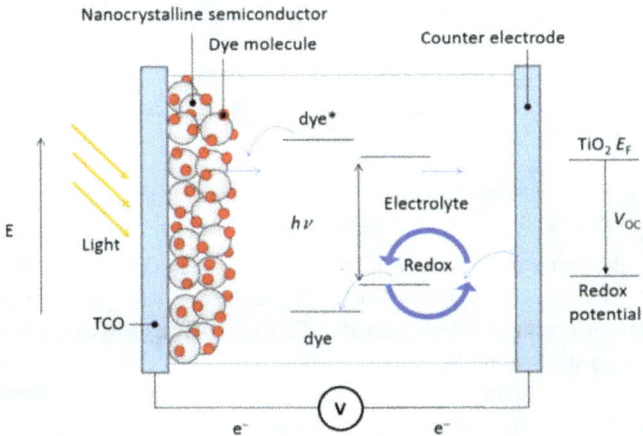

Figure 3: Schematic illustration representing device structure and working principle of a dye-sensitized solar cell adapted from ref [31]. TCO = Transparent Conducting Oxide, $E_f\,TiO_2$ = fermi level of TiO_2, dye* = excited state of dye sensitizer molecule.

During operation, sunlight is absorbed by the dye molecules, exciting them and injecting electrons into the semiconductor's conduction band. These electrons then travel through the semiconductor to the anode and perform useful work at the external load. The electrons are collected by the electrolyte at the counter electrode, which absorbs them to regenerate the dye sensitizer. The dye is regenerated by oxidizing iodide ions to triiodide ions, and platinum acts as a catalyst to accelerate the reduction reaction at the counter electrode as in Fig.4.

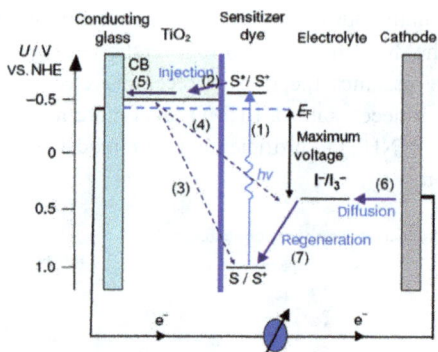

Figure 4: Schematic representation of the operating principle of dye-sensitized solar cells adapted from ref [32].

Interfacial electron transfer kinetics play a crucial role in determining the DSSC's (PCE) [32] and the overall performance of the DSSC can be evaluated based on its PCE, which is calculated using the incident light harvest efficiency (LHE), charge injection and collection efficiencies, as well as the V_{oc} and J_{SC} parameters. By optimizing V_{OC}, J_{SC}, and FF, the device efficiency can be significantly improved under normal operating conditions. Standard illumination at 100 mW/cm^2 and AM 1.5 is commonly used to test the DSSC's performance.

Common degradation factors (as shown in Fig. 5) observed in DSSCs include:

1. Intrusion of Moisture and Oxygen: Moisture and oxygen can penetrate the active area of the cell, leading to corrosion and degradation of the components, especially the dye and semiconductor, which hinders electron transfer and reduces overall efficiency.

2. Electrolyte Sensitivity to UV Light: The electrolyte used in DSSCs may be sensitive to UV light, causing photochemical reactions that degrade the electrolyte, reducing its effectiveness in regenerating the dye and affecting electron transfer.

3. Electrolyte Leakage: Poor sealing or degradation of materials can result in electrolyte leakage, leading to a loss of electrolyte content and reduced efficiency over time.

4. Electrolyte Solvent Evaporation: Under stressful climatic or simulated environmental conditions, the solvent in the electrolyte may evaporate, altering the composition and properties of the electrolyte, and leading to diminished performance.

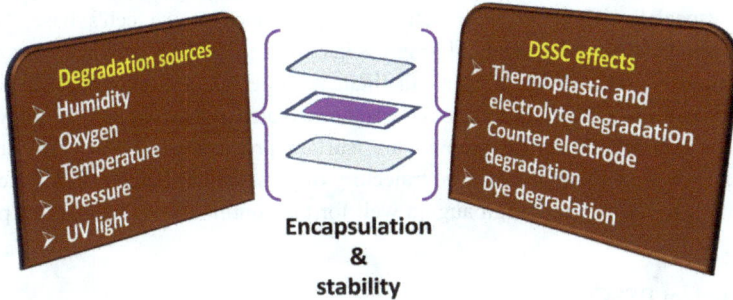

Figure 5: Illustration of the factors that affect DSSC devices and their possible consequences which hinder the photovoltaic performance.

To maintain DSSC performance and prolong their operational life, it is essential to address these degradation factors through improved cell sealing techniques, using UV-resistant electrolytes, and optimizing materials to minimize moisture and oxygen intrusion.

3.1.10 P-DSSCs

In 1999, Lindquist and colleagues pioneered the inception of the first p-Dye Sensitized Solar Cell (p-DSSC), a variant akin to n-DSSCs in most aspects, save for the replacement of TiO_2 with a layer of NiO, a p-type semiconductor. Unlike n-DSSCs where TiO_2 harbors electrons as primary charge carriers, p-DSSCs operate using positive holes (h+) as the principal charge carriers within NiO, leading to reverse electron movement. The choice of NiO as the p-type semiconductor stems from its facile preparation, elevated valence band edge, and substantial band gap (3.6-4.0 eV). However, NiO does entail some drawbacks, including its carcinogenic nature, a contrast to the non-toxic nature of TiO_2. Additionally, higher valence states of NiO, like Ni^{III} and Ni^{IV}, can trigger rapid recombination at the interfaces between dye and semiconductor, as well as between semiconductor and electrolyte, curtailing the lifespan of charge carriers and diminishing overall efficiency. This phenomenon translates to comparatively lower efficiency in p-DSSCs with their n-type counterparts.

The selection of the sensitizer dye stands as a pivotal determinant of DSSC performance. This dye undertakes the responsibility of capturing light photons and transferring their energy to the semiconductor material, thereby optimizing light energy absorption and its subsequent transfer to the semiconductor [33]. In the case of NiO devices, high extinction coefficients are necessary to absorb all incident light, as the film thickness is constrained by the diffusion length. Furthermore, the dye should be proficient in capturing red-NIR

15

photons, especially when the photocathode is positioned at the cell's base. Notably, researchers have achieved substantial progress in elevating p-DSSC efficiency through the formulation of photosensitizers that facilitate charge separation, coupled with the introduction of innovative iodide-free redox intermediates. After a decade of studying the mechanism, electron transfer dynamics, and surface properties, p-DSSC efficiency has surged to 4.1% with tandem cells. The trajectory of persistent research and advancements in material innovation and design augurs well for the continual enhancement of p-DSSC efficiency in the future.

3.1.11 Tandem DSSCs

An effective approach to improve the efficiency of DSSCs involves replacing the single-junction configuration with a multijunction tandem cell. Tandem DSSCs exhibit promising potential in augmenting light harvesting across the entire spectrum by layering multiple dyes with complementary absorption characteristics. Under standardized test conditions, the theoretical upper limit of tandem DSSC efficiency can soar to 43%, a substantial improvement compared to the 30% efficiency observed in conventional DSSCs featuring with only one photoactive dye-sensitized electrode [34].

Tandem DSSCs can be broadly classified into three groups:

1. Stacked arrangements of preassembled DSSC devices.
2. Pairing of dye-sensitized photocathodes with dye-sensitized photoanodes, known as pn-DSSCs.
3. Combinations of complete dye-sensitized solar cells with other solar cell types, forming hybrid structures.

The most straightforward tandem structure involves placing a top stacked on a bottom DSSC. These two cells can be interconnected in either series or parallel configurations, creating tandem DSSCs known as series-connected (ST-DSSC) or parallel-connected (PT-DSSC) setups.

In ST-DSSCs, achieving uniform short circuit photocurrent density values for both the top and bottom cells is crucial, with the aim of generating a higher open-circuit voltage. This approach capitalizes on the higher energy content carried by shorter-wavelength photons compared to their longer-wavelength counterparts.

In contrast, PT-DSSCs offer relatively straightforward matching of V_{oc} values by adjusting the top cell, thereby maximizing J_{sc}. Each of these configurations presents distinct advantages, and the choice of connection strategy depends on specific design objectives and targeted efficiency enhancements.

The implementation of tandem DSSCs empowers researchers to explore diverse combinations of dyes and materials, enabling optimization of light absorption and overall efficiency. This strategic shift brings us closer to the theoretical efficiency limits, unlocking the vast potential harbored within these advanced solar cell configurations.

3.1.12 Advantages

DSSCs offer several advantages, making them an attractive option for renewable energy generation:

1. Cost Effectiveness: DSSCs provide comparable PCE while keeping material and manufacturing costs low.

2. Ambient Temperature Processing: DSSCs can be processed at ambient temperature, enabling the use of a roll-to-roll printing process for mass production, making them cost-effective and scalable.

3. Flexible Thin-Film Structure: DSSCs can be produced as flexible, thin-film structures, allowing for versatile applications and integration into various surfaces and devices.

4. Wide Range of Colors and Transparency: The use of dyes enables the creation of colored and transparent cells, providing aesthetic options and opportunities for integrating solar cells seamlessly into architectural designs and various applications.

5. Advancements in Molecular Engineering: Ongoing developments in molecular engineering have introduced colored and transparent thin films, enhancing aesthetic values while maintaining efficient solar energy conversion.

6. Lightweight and Environmentally Friendly: DSSCs are lightweight and environmentally friendly, contributing to reduced environmental impact compared to conventional energy sources.

7. Recyclability: DSSCs are recyclable, further supporting their eco-friendly nature and sustainable energy solutions.

Overall, DSSCs offer a promising alternative for harnessing solar energy with their cost-effectiveness, versatility, aesthetic appeal, and environmental benefits, making them a key player in the renewable energy landscape.

3.2 Perovskite solar cell

Perovskite solar cells have developed from dye-sensitized solar cells by replacing the molecular dye with a perovskite material. Initially used as a sensitizer in dye-sensitized solid-state devices, perovskite demonstrated its potential by producing a photocurrent with a PCE of around 3–4% on a nanocrystalline TiO_2 surface in 2009 [35]. The PCE was doubled within two years through optimization of the perovskite coating conditions. However, the liquid-based PSC faced stability issues, primarily due to the instant dissolution of perovskite in the liquid electrolyte.

In 2012, a significant breakthrough was achieved with the development of a long-term stable perovskite solar cell, reaching a high efficiency of around 10% by substituting the solid hole conductor with a liquid electrolyte [2]. Over the last decade, perovskite solar cell efficiencies have risen rapidly, reaching impressive values of 18% and even 28% in tandem architecture [36]. PSCs have emerged as a promising photovoltaic device based on organometallic halides. It exhibits several prominent characteristics, including a broad absorption spectrum with a high absorption coefficient, fast carrier mobility, long carrier diffusion lengths, low-cost fabrication processing, and availability of inexpensive raw materials. The exceptional performance of PSCs, with a certified PCE of up to 25.7%, is mainly attributed to the unique physical and chemical properties of the light-harvesting metal halide perovskite, with its structural formula of ABX_3.

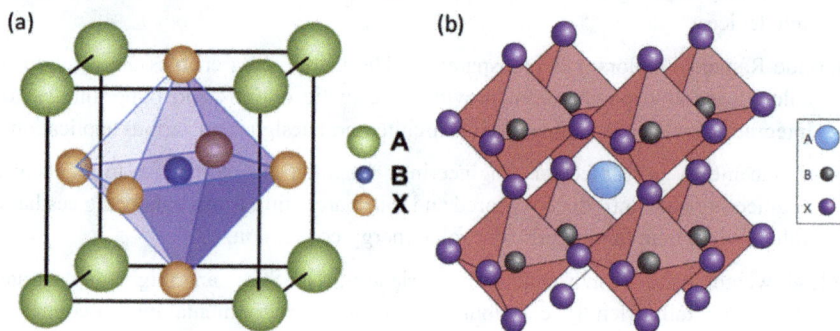

Figure 6: (a) ABX₃ perovskite unit cell (b) extended network structure of perovskites linked via the corner shared octahedral [37].

Perovskite compounds, characterized by the crystal formula ABX_3, where A and B are metal cations of varying sizes and X is an anion (either halogen or oxygen), are named after the Russian mineralogist L.A. Perovski, who discovered their crystal structure in 1839 [37]. In the perovskite structure, A represents a large mono-cation occupying a cubo-octahedral site, B is a smaller di-cation in an octahedral site, and X is an anion (halogen or oxygen) as depicted in Fig 6. Its unique structure of perovskites allows them to accommodate very large cations, including small organic cations, leading to organic-inorganic hybridization. By adjusting the composition and ratio of "A," "B," and "X" ions, the behavior of perovskites can be extensively modified, forming the basis for optimizing the performance of PSCs. Notably, the majority of high-efficiency research is currently focused on three-dimensional (3D) metal halide perovskites [38,39]. However, the long-term stability of 3D perovskites against heat, moisture, and light poses a significant challenge for commercialization, as shown in Fig. 7 [40].

Two-dimensional (2D) perovskites, on the other hand, exhibit natural multi-quantum-well structures and have gained attention as more stable photovoltaic materials [41]. In PSCs, both 3D and 2D perovskites are commonly fabricated as polycrystalline thin films, where the presence of numerous grain boundaries leads to significant non-radiative recombination, ultimately affecting device performance. In comparison, perovskite single crystals exhibit a lower degree of defects, indicating potential for better photovoltaic performance, higher moisture stability, and resistance to ion migration, positioning perovskite solar cells as a robust candidate for commercialization [42].

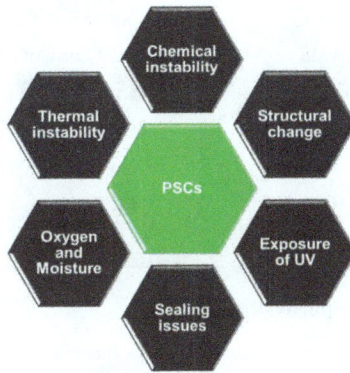

Figure 7: several issues related to stability of PSC's .

3.2.1 Configuration of PSC

A typical PSC device comprises two electrodes, a perovskite light absorber, an n-type electron transport layer (ETL), and a p-type hole transport layer (HTL) [43-47]. The perovskite layer is situated between the ETL and HTL to facilitate efficient charge transport. Charge collection is achieved using outer electrodes, which consist of transparent conducting glass coated with indium tin oxide (ITO) or fluorine-doped tin oxide (FTO) as well as a counter electrode (Au, Ag, or carbon).

Researchers have extensively explored various n-type semiconductor materials as potential ETLs and have identified and examined several HTL options, including polymeric, carbon-based, small molecules, and inorganic compounds. These materials have been listed in Table 1. By optimizing the combination of materials and device architecture, PSCs hold tremendous promise for efficient and affordable solar energy conversion.

Table 1: ETL and HTL materials used for PSCs ref [48]

Sl.No	ETL Materials	HTL Materials
1	TiO_2	Spiro-MeoTAD
2	$SrTiO_3$	PEDOT:PSS
3	ZnO	PCDTBT
4	AZO	DEH, PTAA
5	SnO_3	PANI
6	Al_2O_3	CuSCN
7	WS_2	NiOx
8	CdS	Cu_2O or CuO
9	CdSe	CuI
10	Zn_2SnO_4	$CuGaO_3$
11	PCBM	$CuAlO_2$
12	IGZO	MoO_x
13	WO_3, etc	Cus, MoS_2, etc

3.2.2 Organo metal halide perovskite

Perovskite materials used in PV devices typically have a structure represented as ABX_3, where the A site consists of either organic cations like methylammonium (MA) or formamidinium (FA), or inorganic cations such as Cs or Rb. The B site is occupied by either Pb or Sn, while the X site is filled by halides like I, Br, or Cl [49-52]. These organic-inorganic hybrid perovskites have been extensively studied and demonstrated high performance in PV devices, making them the most widely investigated compounds in this field.

In general, PSCs are constructed with a transparent conducting glass acting as the substrate. On top of this, an electron-selective (n-i-p configuration) or hole-selective (p-i-n configuration) contact is deposited. Next, the perovskite absorbing layer is placed, followed by the hole- or electron-selective layer, and finally, a metal contact is added (Fig. 8). Multijunction PV devices, known as tandems, have also shown great promise. These tandems consist of two band gap-matched absorbers that are stacked monolithically. They offer the advantages of low cost and the potential to exceed the Shockley-Queisser efficiency limit.

3.2.3 Based on charge collection

PSCs can be categorized based on the direction of charge collection (Fig 8):

- **n-i-p Type Devices:** In these devices, electrons are collected from the substrate where the incident light enters. The structure is defined by the layers of Electron Transport Layer (ETL), perovskite absorber layer (i), and Hole Transport Layer (HTL).

- **p-i-n Type Devices:** In this configuration, holes are collected through the substrate. The structure consists of the layers of Hole Transport Layer (HTL), perovskite absorber layer (i), and Electron Transport Layer (ETL).

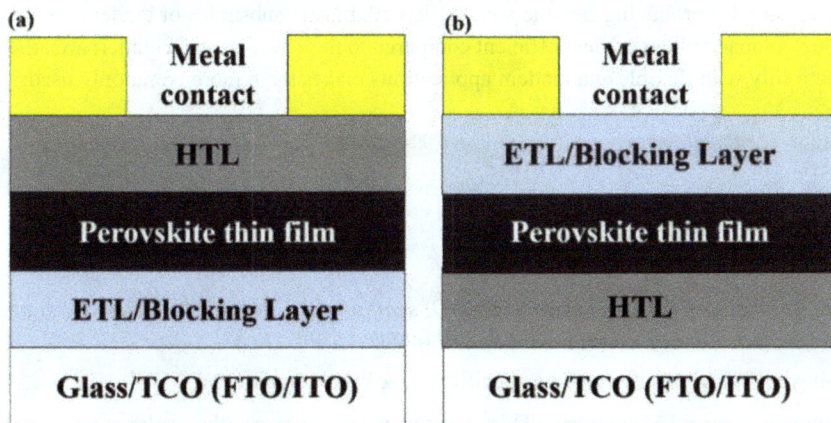

Figure 8: Perovskite solar cells architecture (a) n-i-p configuration: ETL/perovskite/HTL and (b) p-i-n configuration: HTL/perovskite/ETL adapted from ref [53]

3.2.4 Classification of PSCs

Perovskite Solar Cells (PSCs) can be categorized in two distinct manners: structural classification (Fig.9) and charge collection classification.

3.2.5 Structural classification

Mesoscopic Structure: PSCs with a mesoscopic structure resemble DSSCs. These devices feature a thin mesoporous charge transporting layer, similar to DSSCs, but with a more intricate design. In this arrangement, the perovskite layer fills the pores and forms a protective overlayer atop the charge transporting layer [54]. To achieve this, nanoparticles are sintered on a TiO_2 layer to create the mesoporous structure. This architecture is associated with higher processing temperatures. Despite this, it remains the preferred choice for the most efficient single junction PSCs.

Planar Structure: PSCs with a planar structure consist of stacked thin layers, including selective contacts, light-absorbing layers, and window layers [55]. This arrangement is akin to organic photovoltaics or thin-film solar cells such as CdTe, CIGS, and CZTS. Unlike mesoscopic devices, planar devices can be fabricated at lower temperatures, which is advantageous for producing flexible solar cells with plastic substrates or tandem devices. Although planar devices are less efficient compared to their mesoscopic counterparts, their compatibility with flexible and tandem applications makes them more commonly used.

Figure 9: Regular PSC with mesoporous TiO_2 scaffold (left), regular planar (middle) and inverted planar (right) adapted from ref [58]

The above PSCs are represented by specific device configurations as:

Mesoporous n-i-p Structure: This configuration mirrors the solid-state DSSC architecture [56]. It comprises a crystalline organic-inorganic halide perovskite, a mesoporous electron transport layer (often mesoporous TiO_2 with additional compact TiO_2 layers), and a hole transport layer situated between a TCO and a metal contact. The nanoparticles on the TiO_2 layer create the mesoporous structure.

Planar n-i-p Structure: Similar to the mesoporous n-i-p structure, but without the need for mesoporous material. It features a perovskite absorber layer (i) sandwiched between ETL and HTL [57].

Planar p-i-n Structure: This configuration is analogous to the planar n-i-p structure, but the direction of charge collection is reversed. It includes HTL, perovskite absorber layer, and an ETL. Ongoing research and development efforts are focused on addressing these challenges to make PSCs a robust and competitive option for converting solar energy. [58]

3.2.6 Mechanism and operation principle

The operation principle of PSCs is based on the generation and separation of free charge carriers when the perovskite absorber layer is exposed to sunlight [58]. The explanation of the mechanism is as follows:

1. Light Absorption and Charge Generation: When light strikes the perovskite absorber layer, it creates electron-hole pairs (excitons) due to the efficient light harvesting properties of the perovskite material. Unlike in organic solar cells, the small exciton binding energy of the perovskite allows for the immediate generation of free electrons and holes, minimizing energy losses during the process.

2. Charge Transport: The generated free electrons and holes travel through the n-type ETL and p-type HTL, respectively, as they diffuse towards their respective electrodes.

3. Charge Collection: The free electrons reach the cathode by passing through the mesoporous electron transport layer and the external circuit, while the holes diffuse towards the counter electrode direction. At the counter electrode, the holes recombine with electrons, providing the current flow and contributing to the overall power generation (Fig. 10).

The most widely used material for the active layer in PSCs is methylammonium lead iodide perovskite ($MAPbI_3$). It has a direct bandgap of approximately 1.55 eV, making it efficient in absorbing light in the visible range, and the high absorption coefficient ensures a high density of photo-excited charges despite the thin perovskite layer. The optical properties of the perovskite material can be tuned by varying the composition of cations and halogens in the crystal structure. For example, mixing methylammonium lead iodide ($MAPbI_3$) with formamidinium lead iodide ($FAPbI_3$) can enhance light harvesting and current collection, leading to improved photovoltaic performance. Researchers have explored mixed cation and mixed halide perovskite compositions to enhance stability while maintaining high photovoltaic efficiency. For instance, substituting iodide with bromide can improve stability, and the addition of cesium or rubidium cations can achieve even higher efficiency levels.

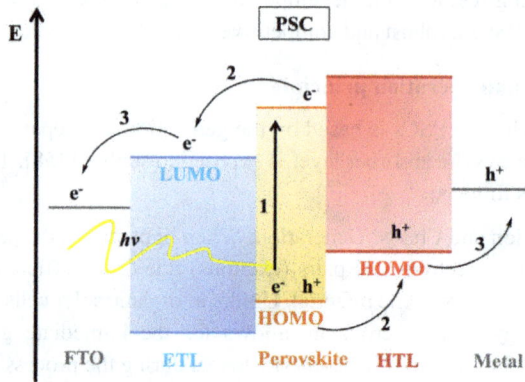

Figure 10: Band diagram and main processes of PSC: 1 Absorption of photon and free charges generation; 2 Charge transport; 3 Charge extraction. adapted from ref [58].

3.2.7 Advantages

PSCs offer numerous advantages that position them as a promising photovoltaic technology for various applications:

1. Lightweight and High Performance: PSCs are lightweight and boast high power conversion efficiencies, allowing them to efficiently convert sunlight into electricity.

2. Flexibility: PSCs can be fabricated as flexible, thin film structures, enabling their integration into various unconventional and curved surfaces

3. Low Cost: The low-cost fabrication processing of PSCs, along with the abundance of raw materials, makes them a cost-effective alternative to traditional solar cells.

4. Broad Application Possibilities: PSCs have a wide range of potential applications, from large-scale solar farms to building-integrated photovoltaics (BIPVs), where solar panels are integrated directly into the building's architecture, thereby serving a dual purpose. Additionally, PSCs can be utilized in small scale applications, like wearable electronics and internet-of-things (IoT) devices.

5. Scalability: PSCs can be easily scaled up for mass production using roll-to-roll processes, which makes them suitable for large-scale manufacturing.

6. Tunability: The optical and electronic properties of perovskite materials can be finely tuned by adjusting the composition, allowing for further performance optimization and customization for specific applications.

7. Efficiency Improvements: Significant advancements have been made in enhancing the efficiency and stability of PSCs, making them competitive with established photovoltaic technologies.

8. Environmental Friendliness: Perovskite materials are environmentally friendly compared to some traditional solar cell materials, making PSCs a more sustainable option.

In a nutshell, the lightweight, high-performance, low-cost, and flexible nature of PSCs, along with their broad application possibilities, position them as a promising candidate for the future of solar energy technology [59-60]. Continued research and development are expected to further enhance their efficiency and stability.

3.3   Organic solar cells

Organic solar cells (OSCs) represent a highly promising and innovative photovoltaic technology with the potential to achieve a good PCE using cost-effective and easily adjustable polymeric or by small molecule organic materials. Moreover, these materials are compatible with existing industrial processes, making large scale production feasible at a low cost. In recent years, significant advancements in the field of OSCs have resulted in a remarkable PCE of 19% [61-62], shedding light on the underlying operation principles and facilitating continuous improvement of the materials and devices. The extensive research efforts have explored various architectures and diverse materials, pushing the boundaries of OSCs.

3.3.1   Classification of OSCs

OSCs exhibit diverse structural configurations, leading to their classification into several categories based on the active layer architecture, as depicted in Fig. 11. The classification of OSCs encompasses single layer, bilayer, bulk heterojunction, ternary, and tandem structures, each offering distinct advantages and challenges. The advancement of these diverse architectures signifies the continual progress and potential of organic solar cell technology.

1. Single Layer OSCs: In the pioneering work of Kearns and Calvin in 1958 (Fig. 11a), the first single layer organic photovoltaic cell was introduced. These early devices faced challenges in achieving efficient charge separation, resulting in a meager maximum output power of 3×10^{-12} W. Due to limited capability in efficiently separating excitons and high rates of electron-hole recombination, the PCE of these single layer devices remained below 0.1%.

2. Bilayer OSCs: A significant stride was made by Tang in 1986 (Fig. 11b) with the introduction of bilayer solar cells. These structures integrated two distinct layers for charge generation and collection. However, challenges such as insufficient charge transport and a lack of interfacial area limited the PCE to around 1%.

3. Bulk Heterojunction (BHJ) OSCs: To tackle the limitations posed by low exciton lifetime and restricted diffusion length, Heeger and collaborators revolutionized the field in 1995 by introducing the concept of Bulk Heterojunction (BHJ) structure (Fig. 11c). This innovative architecture combined acceptor and donor materials, forming an interconnected network with an extensive interfacial area. This facilitated efficient exciton dissociation and enhanced the diffusion distance for exciton separation, resulting in substantial performance improvements. BHJ OSCs have achieved remarkable progress, reaching PCE values of nearly 20%, marking a turning point in OSC technology [63].

4. Ternary and Tandem OSCs: Enhancing OSC performance has been pursued through the development of ternary solar cells (Fig. 11d) and tandem solar cells (Fig. 11e). Ternary OSCs incorporate combinations of donor and acceptor materials, such as donor: acceptor: acceptor (D:A1:A2) or donor:donor:acceptor (D1:D2:A) in the active layer. This approach offers advantages such as energy level tuning, expanded light harvesting spectrum, and morphological adjustments to improve photovoltaic performance. Tandem solar cells introduce an intermediate layer that creates additional charge carriers at distinct electrodes, amplifying sunlight absorption and thereby boosting overall efficiency [65].

5. BHJ OSCs with Interlayers: Bulk Heterojunction (BHJ) OSCs have notably evolved over the last two decades (Table 2). These devices, featured in both standard and inverted architectures (Fig. 12), position a BHJ donor-acceptor layer between a transparent front electrode and a reflective back electrode [66]. Optimal performance necessitates the inclusion of both HTL and ETL acting as selective charge extraction interlayers. These interlayers, preferably amorphous and transparent, contribute to the formation of a smooth layer for effective OSC operation.

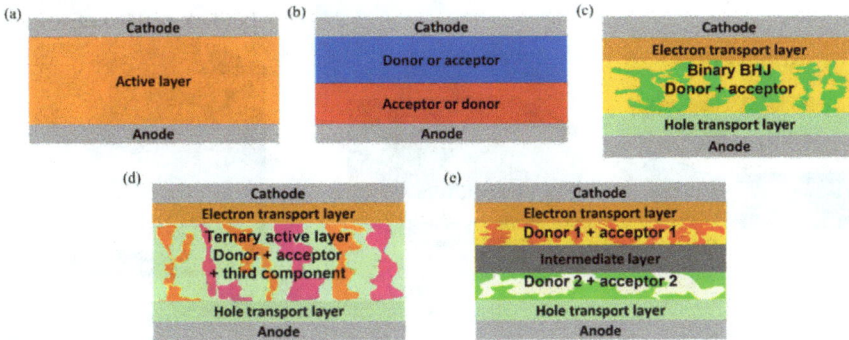

Figure 11: Schematic diagrams of different OSC device structures adapted from ref [64].

3.3.2 Structure and working principle of organic solar cell

3.3.2.1 Structure

The structure of a typical OSC consists of an active layer, charge-transport layers, and electrodes [67-70]. In BHJ OSCs, the active layer is composed of a combination of at least two organic semiconductors serving as both donor and acceptor (Fig.12). This amalgamation can consist of fullerene derivatives, non-fullerene derivatives, or polymers. Analogous to the conduction and valence bands in inorganic semiconductors, the strategic selection of low band gap polymers exhibiting precise HOMO and LUMO energy levels, and certain dyes, are frequently used as donor materials. Conversely, the role of electron acceptors predominantly falls upon fullerene derivatives, perylenediamide, or other n-type substances. The active layer is placed between metal cathodes with low-work function, such as Al, Ag, or Ca, and transparent anode electrodes made of ITO (Indium Tin Oxide). Additionally, ETL made of certain transition metal oxides semiconductors like ZnO, TiO$_2$, and HTL like PEDOT:PSS, MoO$_3$ are incorporated to facilitate the proper extraction of charge carriers [71-74].

Tandem architecture

Regular BHJ

Inverted BHJ

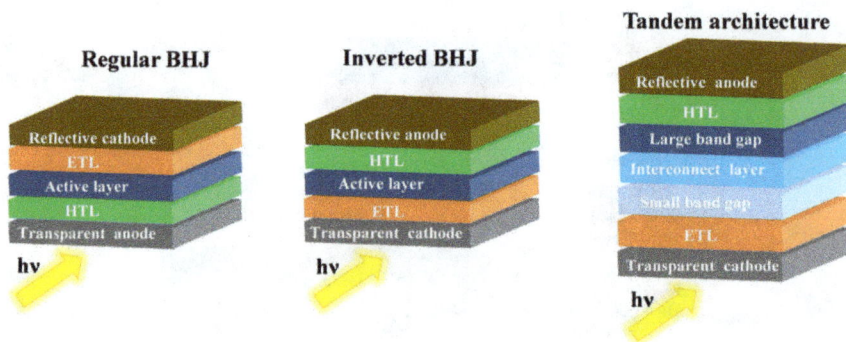

Figure 12: Regular, inverted and Tandem BHJ OSCs adapted from ref [58].

Table 2: Photovoltaic performances of BHJ type binary OSCs devices. ref [64]

Donors	Acceptors	Voc	Jsc	FF	PCE %
PFBDB-T	C8-ITIC	0.94	19.6	72	13.2
PBDB-TF	BTP-eC9	0.841	26.2	78.3	17.3
P3HT	$IC_{70}BA$	0.87	11.53	75.0	7.40
PM6	Y6	0.83	25.63	68.40	14.55
PM6	Y6	0.85	24.45	72.69	15.10
PBDF-NS	Y6	0.728	26.88	72.9	14.26
PBB-T	ITIC-F	0.91	20.9	70	13.3
PBDB-T-SF	IT-4F	0.88	20.50	71.9	13
PBDB-T	IT-M	0.94	17.44	0.735	12.05
PTB7-Th	IEICO-4F	0.71	27.3	66	12.8
PM6	BTP-eC9- 16	0.844	27.78	77.68	18.20
PBQx-TF	eC9-2Cl	0.879	27.2	80.4	19.20
PM6	Y6	0.82	25.2	76.1	15.7
PM6	IPTBO-4Cl	0.893	23.15	72.57	15.00
PBDB-T	ITIC	0.932	17.97	66.2	11.21
Tz6T	eC9-4F	0.863	25.14	70.86	15.38
BTR-Cl	Y6	0.86	24.17	65.5	13.61
PTQ10	MO-IDIC- 2F	0.906	19.87	74.8	13.46
PBDB-T	IDT-EDOT	0.86	21.34	62.0	11.32
D18	Y6	0.859	27.70	76.6	18.22
D18-Cl	Y6	0.87	27.52	75.59	18.15

3.3.2.2 Operational mechanism

The operation of the device involves four crucial stages:

(i) Absorption of the incident photon and generation of excitons,

(ii) Diffusion of excitons,

(iii) Dissociation of excitons, facilitated by the high electron affinity of acceptors, and

(iv) Collection of charges at the electrodes, as shown in Fig. 13.

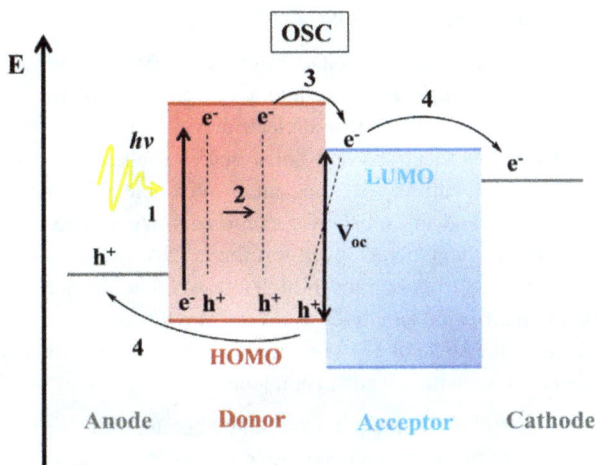

Figure 13: Band diagram and main processes in OSCs: 1 Absorption of photon followed by exciton formation; 2 Exciton diffusion; 3 Charge separation; 4 Charge extraction adapted from ref [58].

3.3.3 Light absorption in OSCs

Light absorption in OSCs starts with the incident photon being absorbed by the active layer, which is a blend of an electron donor (organic p-type material) and an electron acceptor (organic n-type material), after passing through the transparent electrode. This active layer functions as a light harvester while ensuring efficient charge separation. To achieve effective conversion of sunlight into electricity, a photovoltaic device must harvest a significant portion of the solar radiation spectrum. The absorption efficiency depends on the absorption coefficient (α) of the absorbing material and the reflectance of the absorbing surface, as reflected photons are lost. Organic materials typically exhibit high absorption coefficients, allowing for efficient light harvesting with a relatively thin layer (100–200 nm) [71]. However, one limitation of organic molecules is their narrow absorption spectra, with limited coverage in the near-infrared region. Therefore, to enhance the device's performance, it is often necessary to tune the band gap and broaden the absorption by employing appropriate molecular design. One approach to achieve this is by using materials with a high degree of delocalization due to a large π conjugated system, which allows for shifting the donor's absorption to lower energies.

3.3.4 Generation of charge carrier in OSCs

Excitons, formed when an electron is excited from the HOMO to the LUMO, create a bound electron-hole pair with a high columbic interaction energy known as exciton binding energy (E_b). Organic materials with low dielectric constant say ($\varepsilon_r \sim 2\text{-}3$), the strong coulombic attraction between electrons and holes makes it challenging for excitons to dissociate. To produce photocurrent, excitons must overcome the binding energy and dissociate into free electrons and holes. In OSCs, exciton dissociation occurs at the junction where p-type and n-type semiconductors meet, forming a D/A interface. The migration of excitons is a diffusive process from regions of high to low exciton concentration, which is dependent on the exciton diffusion length (DL) and the film thickness. Typically in organic semiconductors, DL is in the range of 1–10 nm, limiting the absorbing layer thickness and necessitating exciton creation within its diffusion length.

At the D/A interface in a Bulk BHJ, exciton dissociation begins with electron transfer from the exciton to the LUMO of the acceptor, resulting in D+/A- at the interface namely charge transfer state (CT). If the distance between the electron and hole exceeds the Coulomb capture radius, full charge separation occurs, and a charge-separated state (CS) with free electrons and holes is formed. The generated charge carriers are then transferred to the electrodes via the donor and acceptor materials, and charge transport occurs within nanoseconds. The charge transfer efficiency (ηCT) represents the ratio of transferred charges at the D/A interface to the total excitons. Minimizing recombination losses is crucial for achieving high performance OSCs, as recombining charge carriers impede their movement towards the hole and electron contacts [75]. High short-circuit current density in OSCs can be achieved by using donor materials with small optical bandgaps (E_g), while the open-circuit voltage is determined by the energy gap between the HOMO of the donor and LUMO of the acceptor materials. Electrically well-aligned material structures facilitate the movement of charge carriers, reducing recombination in active layers and increasing the FF of OSCs [76].

3.3.5 Advantages

OSCs harness organic materials to generate electrons for power generation, and the abundant availability of these organic constituents helps in reducing manufacturing costs. The use of organic BHJs has shown promising applications in transparent, flexible photovoltaic devices, as well as indoor photovoltaic devices, owing to their unique material properties. Significant progress has been made in enhancing the performance of OSCs, with a remarkable PCE of up to 19%. This progress is attributed to the development of novel acceptor and donor materials, the use of interfacial materials to improve charge-carrier collection [77-80], and the optimization of phase-separation morphology within the active layer. These advancements have paved the way for the widespread adoption and further exploration of OSC technology.

4.   Applications of third-generation solar cells

The progression of PV solar cells across different generations has been driven by the pursuit of cost effectiveness in photovoltaic technology [81]. Factors such as financial considerations, environmental impacts, and net accessible energy play pivotal roles in determining the performance of PV systems. OSCs have emerged as promising solutions for eco-friendly, thin, and flexible low-cost photovoltaic devices, finding applications in various areas such as IoT sensors and wearable electronics. The integration of indoor solar cells can significantly impact the ecosystem of IoTs, encompassing portable communication devices, actuators, and remote and decentralized sensors. Smart IoT sensors, which require minimal power for specific applications can greatly benefit from indoor power collection systems. The adaptability of DSSCs as fenestration in buildings gives them a fundamental advantage over other current technologies. The transparency of DSSCs makes them suitable for outdoor applications, such as windows in building-integrated photovoltaic systems [82-84]. Additionally, DSSCs exhibit better performance under lower light intensities, making them an excellent choice for indoor applications like windows and sunroofs. Their demonstrated benefits have been observed in medical and sports gear, cameras, security sensors, and wireless sensor network applications, among others.

5. Summary

This chapter focuses on three different third-generation photovoltaic solar cells and their significance. DSSCs offer consistent power output under various illumination situations, including fluorescent and LED lights. Their manufacturing process has been simplified, resulting in an improved PCE from 7% to 14%. PSCs are a relatively newer class of photovoltaic devices that have experienced a remarkable rise in PCE and emerged as highly efficient technology. The high efficiency of PSCs is primarily achieved through the optimization of individual layers in the device. On the other hand, organic tandem solar cells require a more complicated production procedure, which has reduced their efficiency to 10%. Both OSCs and PSCs are different in the mechanism of charge generation due to their distinct active layer materials, namely organic semiconductors and hybrid organic-inorganic perovskites. Despite their differences, they share some similarities in materials processing, with standard strategies developed for OSCs currently being employed in PSCs.

In summary, third-generation solar cells encompass a diverse range of technologies that aim to push the boundaries of solar energy conversion efficiency and explore new materials and mechanisms for solar cell operation. There are still many directions open to research in electrodes, photosensitizers, electrolytes, active layer and donor/acceptor polymers, which could eventually lead to more efficient, affordable, and environmentally friendly solar energy solutions.**References**

[1] D.M. Chapin, C.S. Fuller, G. L. Pearson. J.Appl. Phys. 25, 676, (1954). https://doi.org/10.1063/1.1721711

[2] C.E. Frittis, Am. J. Sci. 26, 465, (1883).

[3] W. Smith, Nature, 7, 303, (1873). https://doi.org/10.1038/007303c0

[4] D. Neuhaus, M. Adolf, Adv. Optoelectron. 024521, 1, (2007).

[5] M. Gul, Y. Kotak, T. Muneer, Energy Explor. Exploit. 34, 485, (2016). https://doi.org/10.1177/0144598716650552

[6] C. Battaglia, A. Cuevas, S. DeWolf, Energy Environ. Sci. 9, 1552, (2016). https://doi.org/10.1039/C5EE03380B

[7] B. Parida, S. Iniyan, R. Goic, Renew. Sustain. Energy Rev. 15, 1625, (2011.) https://doi.org/10.1016/j.rser.2010.11.032

[8] W. Shockley, H.J. Queisser, J. Appl. Phys. 32, 510, (1961). https://doi.org/10.1063/1.1736034

[9] K Pal, KB Thapa, A Bhaduri, Adv. Sci. Engg and Med., 10, 645, (2018). https://doi.org/10.1166/asem.2018.2225

[10] S Siebentritt, Thin Solid Films, 535, 1, (2013). https://doi.org/10.1016/j.tsf.2012.12.089

[11] Z. Hens, I. Moreels, J. Mater. Chem, 22, 10406, (2012). https://doi.org/10.1039/c2jm30760j

[12] Y. Hishikawa, E.D. Dunlop, D.H. Levi, M.A. Green, J. Hohl, E. Masahiro, Y. Anita, W.Y.H. Baillie, Prog. Photovolt. Res. Appl. 27, 565, (2019).

[13] B. O'regan, M. Grätzel, Nature, 353, 737, (1991). https://doi.org/10.1038/353737a0

[14] A. Hagfeldt, G. Boschloo, L. Sun, L. Kloo, H. Pettersson, Chem. Rev. 110, 6595, (2010). https://doi.org/10.1021/cr900356p

[15] F. Babar, U. Mehmood, H. Asghar, M. H. Mehdi, A. Ul Haq Khan, H. Khalid, N. Huda, Z. Fatima, Renew Sustain Energy Rev , 129, 109919, (2020). https://doi.org/10.1016/j.rser.2020.109919

[16] I. Joseph, H. Louis, T. O. Unimuke, I. S. Etim, M. M. Orosun, J. Odey, Appl. Sol. Energy., 56, 334, (2020). https://doi.org/10.3103/S0003701X20050072

[17] J. Gong, J. Liang, K. Sumathy, Renew Sustain Energy Rev, 16, 5848-60, (2012). https://doi.org/10.1016/j.rser.2012.04.044

[18] A. Hagfeldt, G. Boschloo, L. Sun, L. Kloo, H. Pettersson, Chem Rev, 110, 6595, (2010). https://doi.org/10.1021/cr900356p

[19] N. A Karim, U Mehmood , HF Zahid, T. Asif, Sol Energy, 185, 165, (2019). https://doi.org/10.1016/j.solener.2019.04.057

[20] D. Sengupta, P. Das, B. Mondal, K. Mukherjee, Renew Sustain Energy Rev, 60, 356, (2016). https://doi.org/10.1016/j.rser.2016.01.104

[21] K. Sayama, H. Sugihara, H. Arakawa, Chem Mater, 10, 3825, (1998). https://doi.org/10.1021/cm9801111

[22] M. K. Nazeeruddin, P. Liska, J. Moser, N. Vlachopoulos, M. Gr€atzel, Helv , Chim Acta, 73, 1788, (1990). https://doi.org/10.1002/hlca.19900730624

[23] M. K. Nazeeruddin, P. Pechy, M. Gr€atzel, Chem Commun, 6, 1705, (1997). https://doi.org/10.1039/a703277c

[24] M. R. Narayan, Renew Sustain Energy Rev, 16, 208, (2012).

[25] D. Kuang, C. Klein, S. Ito, J.E. Moser, R. Humphry-Baker, N. Evans, F. Duriaux, C. Gr€atzel,S.M. Zakeeruddin, M. Gr€atzel , Adv Mater, 19,1133, (2007). https://doi.org/10.1002/adma.200602172

[26] H. Kusama, H. Arakawa, Energy Mater Sol Cells, 85, 333, (2005). https://doi.org/10.1016/j.solmat.2004.05.003

[27] A.F. Nogueira, C. Longo, M.A. De Paoli, Coord Chem Rev, 248, 1455, (2004). https://doi.org/10.1016/j.ccr.2004.05.018

[28] SM Zakeeruddin, M Gratzel, Adv. Funct. Mater. , 19, 2187, (2009). https://doi.org/10.1002/adfm.200900390

[29] J. Wu, Z. Lan, S. Hao, P. Li, J. Lin, M. Huang, L. Fang, Y. Huang, Pure Appl Chem, 80, 2241, (2008). https://doi.org/10.1351/pac200880112241

[30] Y-H Wei, M-C Tsai, C-CM Ma, H-C Wu, F-G Tseng, C-H Tsai , C-K Hsieh, Nanoscale. Res. Lett., 10, 467, (2015).

[31] J. M. Cole, U.F.J. Mayer, Langmuir, 38, 871, (2022). https://doi.org/10.1021/acs.langmuir.1c02165

[32] V. Rondán-Gómez, I. Montoya De Los Santos, D. Seuret-Jiménez, F. Ayala-Mato, A. Zamudio-Lara, T. Robles-Bonilla, M. Courel, Appl. Phys. A., 125, 836 (2019). https://doi.org/10.1007/s00339-019-3116-5

[33] [JR Durrant, SA Haque, E Palomares, Coord. Chem. Rev, 248, 1247 (2004). https://doi.org/10.1016/j.ccr.2004.03.014

[34] SCT Lau, J Dayou, CS Sipaut, RF Mansa, Int J Renew Energy Resour, 4, 665, (2014).

[35] A. Kojima, K. Teshima, Y. Shirai, T. Miyasaka. J. Am Chem Soc, 131 ,6050, (2009). https://doi.org/10.1021/ja809598r

[36] H. S. Kim, C. R. Lee, J. H. Im, K. B. Lee, T. Moehl, A. Marchioro and N. G. Park, , Sci. Rep., 2, 591, (2012).

[37] Md Moniruddin, B Ilyassov, X Zhao, E Smith, T Serikov, N Ibrayev, R Asmatulu, N Nuraje, Mater. Today Energy, 7, 246-259, (2018). https://doi.org/10.1016/j.mtener.2017.10.005

[38] JY Kim, JW Lee, HS Jung, H Shin, NG Park, Chem Rev, 120, 7867, (2020). https://doi.org/10.1021/acs.chemrev.0c00107

[39] J Xiao, J Shi, D Li, Q Meng, Sci China Chem, 58, 221, (2015). https://doi.org/10.1007/s11426-014-5289-2

[40] T. A. Chowdhury, Md. A. B. Zafar, Md. S. Islam, M. Shahinuzzaman, M. A. Islam, M. U. Khandaker, RSC Adv., 13, 1787, (2023). https://doi.org/10.1039/D2RA05903G

[41] J Yan, w Qiu, G Wu, P Heremans, H Chen, J Mater Chem A, 6, 11063, (2018) https://doi.org/10.1039/C8TA02288G

[42] MA Green, A Ho-Baillie, HJ Snaith. Nat Photon, 8, 506, (2014). https://doi.org/10.1038/nphoton.2014.134

[43] MA Haque, J Troughton, D Baran. Adv Energy Mater, 10, 1902762, (2020). https://doi.org/10.1002/aenm.201902762

[44] B. Chen, M. Yang, S. Priya, K. Zhu, J. Phys. Chem. Lett. 7, 905, (2016). https://doi.org/10.1021/acs.jpclett.6b00215

[45] S. De Wolf, J. Holovsky, S.J. Moon, P. Löper, B. Niesen, M. Ledinsky, F.J. Haug, J.H. Yum, C. Ballif, J. Phys. Chem. Lett. 5, 1035, (2014). https://doi.org/10.1021/jz500279b

[46] R. Sheng, A. Ho-Baillie, S. Huang, S. Chen, X. Wen, X. Hao, M.A. Green, J. Phys. Chem. C, 119, 3545, (2015). https://doi.org/10.1021/jp512936z

[47] J.H. Noh, S.H. Im, J.H. Heo, T.N. Mandal, S.I. Seok, Nano Lett. 13, 1764, (2013). https://doi.org/10.1021/nl400349b

[48] T. A. Chowdhury, Md. A. B. Zafar, Md. S. Islam, M. Shahinuzzaman, M. A. Islam, M. U. Khandaker, RSC Adv., 13, 1787, (2023). https://doi.org/10.1039/D2RA05903G

[49] O. Knop, R.E. Wasylishen, M.A. White, T.S. Cameron, M.J.M. Van Oort, Can. J. Chem. 68, 412, (1990). https://doi.org/10.1139/v90-063

[50] G.E. Eperon, S.D. Stranks, C. Menelaou, M.B. Johnston, L.M. Herz, H.J. Snaith, Energy Environ. Sci. 7, 982, (2014). https://doi.org/10.1039/c3ee43822h

[51] N. Pellet, P. Gao, G. Gregori, T.Y. Yang, M.K. Nazeeruddin, J. Maier, M. Grtäzel, Angew. Chemie - Int. Ed. 53, 3151, (2014). https://doi.org/10.1002/anie.201309361

[52] D. Bi, W. Tress, M.I. Dar, P. Gao, J. Luo, C. Renevier, K. Schenk, A. Abate, F. Giordano, J.-P. Correa Baena, J.-D. Decoppet, S.M. Zakeeruddin, M.K. Nazeeruddin, M. Gra tzel, A. Hagfeldt, Sci. Adv. 2, 1501170, (2016).

[53] S. Nair, S. B. Patel, J. V. Gohel, Mater. Today Energy 17, 100449, (2020). https://doi.org/10.1016/j.mtener.2020.100449

[54] C. Yi, X. Li, J. Luo, S.M. Zakeeruddin, M. Grätzel, D.P. McMeekin, G. Sadoughi, W. Rehman, G.E. Eperon, M. Saliba, M.T. Horantner, A. Haghighirad, N. Sakai, L. Korte, B. Rech, M.B. Johnston, L.M. Herz, H.J. Snaith, Science (80), 351, 151-155, (2016). https://doi.org/10.1126/science.aad5845

[55] M. Saliba, T. Matsui, K. Domanski, J.Y. Seo, A. Ummadisingu, S.M. Zakeeruddin, J.P. Correa-Baena, W.R. Tress, A. Abate, A. Hagfeldt, M. Grätzel, Science 354, 206, (2016). https://doi.org/10.1126/science.aah5557

[56] M. Saliba, T. Matsui, J.-Y. Seo, K. Domanski, J.-P. Correa-Baena, M.K. Nazeeruddin, S.M. Zakeeruddin, W. Tress, A. Abate, A. Hagfeldt, M. Grätzel, Energy Environ. Sci. 9, 1989, (2016). https://doi.org/10.1039/C5EE03874J

[57] A. Miyata, A. Mitioglu, P. Plochocka, O. Portugall, J.T.-W. Wang, S.D. Stranks, H. J. Snaith, R.J. Nicholas, Nat. Phys. 11, 582, (2015). https://doi.org/10.1038/nphys3357

[58] N. Marinova, S. Valero, J. L. Delgado, Jour. of Colloid and Interf. Sci, 488, 373, (2017). https://doi.org/10.1016/j.jcis.2016.11.021

[59] E. Edri, S. Kirmayer, S. Mukhopadhyay, K. Gartsman, G. Hodes, D. Cahen, Nat. Commun. 5, 3461, (2014). https://doi.org/10.1038/ncomms4461

[60] Q. Lin, A. Armin, R.C.R. Nagiri, P.L. Burn, P. Meredith, Nat. Photonics 9, 106, (2014). https://doi.org/10.1038/nphoton.2014.284

[61] Y. Cui, Y. Xu, H. Yao, P. Bi, L. Hong, J. Zhang, Y. Zu, T. Zhang, J. Qin, J. Ren, Z. Chen, C. He, X. Hao, Z. Wei, J. Hou, Adv. Mater. 33, 2102420, (2021). https://doi.org/10.1002/adma.202102420

[62] H. Kallmann, M. Pope, Photovoltaic effect in organic crystals, J. Chem. Phys. 30, 585, (1959). https://doi.org/10.1063/1.1729992

[63] C.W. Tang, Two-layer organic photovoltaic cell, Appl. Phys. Lett. 48, 183, (1986). https://doi.org/10.1063/1.96937

[64] W.B. Tarique and A. Uddin , Mater. Sci. in Semicond. 163, 107541, (2023). https://doi.org/10.1016/j.mssp.2023.107541

[65] G. Yu, J. Gao, J.C. Hummelen, F. Wudl, A.J. Heeger, Science, 270, 1789, (1995). https://doi.org/10.1126/science.270.5243.1789

[66] J. Fu, P.W.K. Fong, H. Liu, C.-S. Huang, X. Lu, S. Lu, M. Abdelsamie, T. Kodalle, C. M. Sutter-Fella, Y. Yang, G. Li, Nat. Commun. 14, 1760, (2023).

[67] L. Meng, Y. Zhang, X. Wan, C. Li, X. Zhang, Y. Wang, X. Ke, Z. Xiao, L. Ding, R. Xia, H.-L. Yip, Y. Cao, Y. Chen, Science , 361, 1094, (2018). https://doi.org/10.1126/science.aat2612

[68] X. Xu, Y. Li, Q. Peng, Nano Select 1, 30, (2020). https://doi.org/10.1002/nano.202000012

[69] X. Guo, D. Li, Y. Zhang, M. Jan, J. Xu, Z. Wang, B. Li, S. Xiong, Y. Li, F. Liu, J. Tang, C. Duan, M. Fahlman, Q. Bao, Org. Electron. 71, 65, (2019). https://doi.org/10.1016/j.orgel.2019.05.004

[70] Z. Hu, F. Zhang, Q. An, M. Zhang, X. Ma, J. Wang, J. Zhang, J. Wang, ACS Energy Lett. 3, 555, (2018). https://doi.org/10.1021/acsenergylett.8b00100

[71] X. Xu, L. Yu, H. Meng, L. Dai, H. Yan, R. Li, Q. Peng, Adv. Funct. Mater. 32, 2108797, (2022).

[72] X. Huang, B. Sun, Y. Li, C. Jiang, D. Fan, J. Fan, S.R. Forrest, Appl. Phys. Lett. 116, 153501, (2020). https://doi.org/10.1063/5.0005172

[73] H.-L. Yip, A.K.-Y. Jen, Energy Environ. Sci. 5, 5994, (2012). https://doi.org/10.1039/c2ee02806a

[74] Y. Wu, H. Bai, Z. Wang, P. Cheng, S. Zhu, Y. Wang, W. Ma, X. Zhan, Energy Environ. Sci. 8, 3215, (2015). https://doi.org/10.1039/C5EE02477C

[75] R. Sorrentino, E. Kozma, S. Luzzati, R. Po, Energy Environ. Sci. 14, 180, (2021). https://doi.org/10.1039/D0EE02503H

[76] Y. Lin, J. Wang, Z.-G. Zhang, H. Bai, Y. Li, D. Zhu, X. Zhan, Adv. Mater. 27, 1170, (2015). https://doi.org/10.1002/adma.201404317

[77] P. Li, Z.-H. Lu, Small Science 1, 2000015, (2021).

[78] H. Zhang, Y. Li, X. Zhang, Y. Zhang, H. Zhou, Mater. Chem. Front. 4, 2863, (2020). https://doi.org/10.1039/D0QM00398K

[79] J.L. Brédas, J.E. Norton, J. Cornil, V. Coropceanu, Acc. Chem. Res. 42, 1691, (2009). https://doi.org/10.1021/ar900099h

[80] K.A. Mazzio, C.K. Luscombe, Chem. Soc. Rev. 44, 78, (2015). https://doi.org/10.1039/C4CS00227J

[81] K. Feron, W.J. Belcher, C.J. Fell, P.C. Dastoor, Int. J. Mol. Sci. 13, 17019, (2012). https://doi.org/10.3390/ijms131217019

[82] A. Ghosh, P. Selvaraj, S. Sundaram, T.K. Mallick, Sol. Energy 163, 537, (2018). https://doi.org/10.1016/j.solener.2018.02.021

[83] E. Mirabi, F. A. Abarghuie ,R. Arazi, Clean Energy, 5, 505, (2021). https://doi.org/10.1093/ce/zkab031

[84] K. Yoshikawa, H. Kawasaki, W. Yoshida, T. Irie, K. Konishi, K. Nakano, T. Uto, D. Adachi, M. Kanematsu, H. Uzu and K. Yamamoto, Nat. Energy. 2, 17032, (2017). https://doi.org/10.1038/nenergy.2017.32

Chapter 2

Recent Developments in Materials for Dye-Sensitized Solar Cells

Santhosh Narendhiran, Manoj Balachandran

Department of Physics and Electronics, Christ (Deemed to be University), Bengaluru-560029, Karnataka, India

santhosh.10409@gmail.com; manoj.b@christuniversity.in

Abstract

Over the past few years, a variety of photovoltaic devices have been developed to serve a diverse range of applications. These devices include inorganic, organic, and hybrid solar cells. Despite the high conversion efficiency of silicon-based solar cells, their practicality is constrained by the expensive module costs and the intricate manufacturing process involved. Among them, dye-sensitized solar cells have emerging technology to power a portable electronic device and work with both outdoor and indoor light, reducing the energy demand in the present scenario. This chapter explains the working mechanism and components of DSSC materials and provides insights into recent developments.

Keywords

Solar Cells, Light-Sensitive Dye, Energy, Photovoltaic Cells

Contents

Third Generation Photovoltaic Technology　　　　　　　　　Materials Research Forum LLC
Materials Research Foundations 163 (2024) 38-51　　　　https://doi.org/10.21741/9781644903032-2

1.　Introduction

As the world's population expands on a daily/constant basis, so are energy demands, and greenhouse gas emissions and climate change are taking notice. Researchers are searching alternative green approaches. Solar energy offers a greener energy source, and it is available abundantly, noiseless and greenhouse gas-free with vast amounts of energy to turned into electricity to meet the demands of world energy consumption. For the past fifty years, photovoltaic technologies, particularly those based on crystalline silicon, have dominated the market, capturing a staggering 90% share [1]. This remarkable achievement can be attributed to their exceptional ability to convert solar energy into electricity when exposed to sunlight, their unwavering performance across diverse climate conditions, and their abundant availability. Si-PV has several drawbacks, such as a costly production procedure and inadequate performance in low-light conditions, particularly for indoor and building integration purposes [2,3].

Third-generation photovoltaic technologies are designated as evolving solar cells based on solution-processable technology, which includes dye-sensitized, organic, quantum-bot, polymer and perovskite solar cells. These solar cells have been manufactured using low-cost, simple, widespread, and scalable production processes. However, these PV generations have a research gap of stability issues and these technologies are still under research/laboratory stage only. In addition, DSSCs have unique advantages over third-generation PV technologies in terms of their specific applications. Because of its easy viability such as cost-effective, indoor light performance, choice of colour selections, building integration and mechanical durability [4,5]. In the year 1991, a significant breakthrough was made by Oregan and Gratzel as they successfully developed a novel nanocrystalline solar cells shows a power conversion efficiency of 7.1% [6]. This ground-breaking achievement not only highlighted the ingenuity of their work but also sparked immense curiosity and enthusiasm within the scientific community, prompting further exploration and investigation into the performance capabilities of these innovative devices known as DSSCs. They have a number of advantages, including their impressive efficiency as well as their simple structure and cost-effectiveness, which makes them a promising research and development prospect.

Third Generation Photovoltaic Technology   Materials Research Forum LLC
Materials Research Foundations 163 (2024) 38-51   https://doi.org/10.21741/9781644903032-2

2.   Working Mechanism of DSSC

The functioning of DSSCs deviates from that of conventional solar cells and resembles the natural process of photosynthesis, as it involves distinct substances facilitating the absorption of light and the movement of charge carriers. The process involves several steps. First, a sensitizer dye is excited by light, causing photoexcitation. Next, the excited dye transfers an electron to the conduction band of a semiconductor metal oxide. This electron then flows to the counter electrode (CE) through an external circuit. At the CE interface, an oxidized electrolyte is reduced. The oxidised dye molecule is finally reduced by an electrolyte.

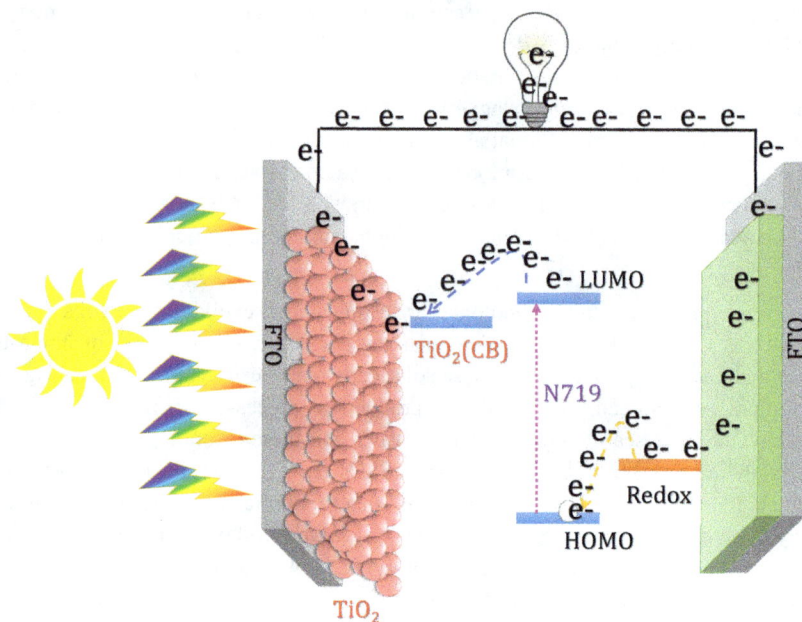

Figure 1. Schematic diagram of DSSC working mechanism

When two chemical reactions competing with each other occur simultaneously in DSSCs, it becomes evident that there are two separate routes for recombination, which involve the oxidation of both dye molecules and the redox electrolyte [7]. Asbury et al., found that the electron transfer time at dye/semiconductor photoanode is $\sim 10^{-15}$ s and dye/electrolyte is 10^{-8} s [8]. This indicates that the reduction of electrolyte is much slower than the charge transfer between photoanode and sensitizer. The dynamics of recombination during the electrolyte reduction process differ from those during the injection process at the TiO_2/dye interface, suggesting that recombination is less likely to occur during the quicker injection process.

3.   Components of DSSC

The DSSC device is made up of various components that work together to generate electricity. These components include a transparent conducting oxide (TCO) glass substrate, a working electrode (photoanode), a sensitizer that absorbs sunlight, an electrolyte that helps transfer electrons to the dye, and a counter electrode that carries electrons from the external circuit and facilitates effective electrolyte reduction for the DSSC to function properly. (i) Transparent conducting glass substrate are the base for the fabrication of DSSC with sheet resistance of less than 10 ohms with more than light transmission of 80%. (ii) Photoanode are constructed from semiconductor nanostructures, such as nanospheres, nanorods, nanowires, nanoleaves etc. (iii) Dye molecules are responsible for the carrier generation. (iv) The electrolyte in DSSCs is a crucial compound that has a significant impact. Its role is essential as it facilitates the movement of charges or ions between the counter electrode and photoanode by utilizing the redox mediator present in the electrolyte. (v) Counter electrode plays a crucial role in the process' completion because it receives electrons and makes it easier to reduce the oxidised redox pair on its surface [9].

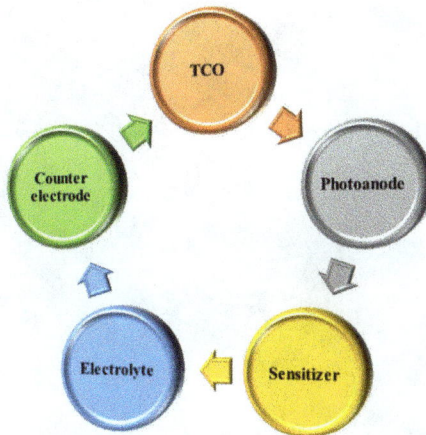

Figure 2. Components of dye-sensitized solar cells

3.1 Recent advancement of photoanode

When building photoanodes for DSSCs, nanocrystalline semiconducting coatings are essential (summarized in Figure 2). Acting as a platform, the photoanode fulfils two important functions: it supports the sensitizer and facilitates the transfer of photo-excited electrons to an external circuit. To accomplish these tasks effectively, a significant surface area is crucial to accommodate a high dye loading. Additionally, to maximize the efficiency of electron collection, a rapid transit rate for charge is indispensable. It is the combination of these two characteristics that distinguishes an ideal photoanode. The photoanode is usually composed of a film with an average thickness of about 10 μm, which is made up of randomly distributed spherical TiO_2 nanoparticles in the form of a structure with three dimensions [10,11]. These nanoparticles have a large surface area, enabling them to hold a high amount of dye. However, the disordered nature of the nanoparticle network, with multiple grain boundaries, hinders the movement of electrons, leading to delays in their transit and recombination. This limitation greatly affects the overall efficiency of DSSCs. Consequently, researchers are actively seeking alternative nanostructured materials and morphologies for the photoanode to overcome these inherent issues.

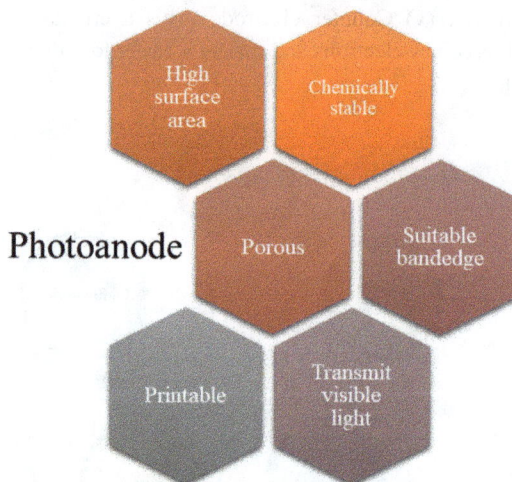

Figure 3. Characteristics of efficient Photoanode in DSSC

3.1.1 Modified Photoanode

To develop the efficient DSSC, photoanode nanostructure will be modified utilizing hydrothermal, electrospinning, and electrochemical anodization processes. These approaches have been instrumental in fabricating various nanostructured semiconductor photoanodes, encompassing nanorods, mesoporous structures, nanotubes, nanosheets, and 3D hierarchical architectures. Various ways for modifying the photoanode are used to improve the efficiency of solar cells. These techniques include the introduction of metal ions through a process known as doping, the addition of noble metals or carbon allotropes for decoration, the utilization of insulating metal oxides, and the incorporation of up/down conversion materials [12,13].

- Ion doping is a commonly employed technique for modifying the valance band (VB) or conduction band (CB) of the semiconductor photoanode. This process involves introducing ions into the photoanode material, which helps reduce recombination resistance and improves the lifespan of electrons in DSSC.

- Noble metals like silver (Ag) or gold (Au) nanostructures for decoration has the effect of creating surface plasmons, which can concentrate incoming light and make it travel a longer distance. These plasmonic effect enhance the light harvesting ability of photoanode, which converts light energy into electricity.

- The utilization of carbon allotropes like graphene quantum dots, carbon quantum dots, graphene, and reduced graphene oxide in the process of decorating enhances the quality of the photoanode by minimizing the presence of defects. As a result, the overall charge collection and transport within the photoanode are significantly improved [14,15].

- The photoanode surface is modified by introducing insulating or semiconducting materials such as Al_2O_3, MgO, $SrCO_3$, $SrTiO_3$, and SnO_2, which act as reduced charge recombination in the heterostructure photoanode [16,17].

- To enhance the current density of DSSC, either down-conversion or up-conversion materials have been utilized. This kind of material absorbs either lower or higher energy and emits visible light, which can be absorbed by the sensitizer and generate more charge carriers. This conversion allows for a more efficient utilization of the light spectrum, ultimately leading to an increase in the overall performance and power output of the DSSC [18].

3.2 Development in sensitizer

In order to effectively carry out their intended function, sensitizers must meet several crucial criteria. Among the various dyes available, Ru-bipyridyl compounds like N719, N3, and black dye have been found to exhibit excellent performance. These dyes have shown superior properties in terms of their ability to convert light into electricity efficiently. The development of novel and innovative dyes has the potential to significantly improve the overall efficiency and stability of dye-sensitized solar cells. Therefore, ongoing research in this field aims to discover new dyes that surpass the performance of existing ones, leading to advancements in solar cell technology. These criteria cover several important factors that a dye must possess in order to be effective in dye-sensitized solar cells. Firstly, the dye should be able to easily absorb light in the visible to near-infrared range. Additionally, it should be chemically stable in both the highest occupied molecular orbital (HOMO) and lowest unoccupied molecular orbital (LUMO) states. Furthermore, it should facilitate the regeneration of the dye from the electrolyte and demonstrate strong resistance to degradation under light exposure (photostability). Lastly, it should have good solubility to prevent recombination of charges [19]. There are several key characteristics that these dyes possess, including those that are based on metal complexes, those that are metal-free, and those that are derived from porphyrin or natural sources. Metal complex dyes, particularly those that utilize ruthenium, have proven to be highly effective in terms of light absorption and conversion. However, there are several disadvantages associated with the use of metal-based sensitizers in DSSCs. These disadvantages include their limited availability, challenges in the purification process, potential harm to the environment, and their relatively low molar extinction coefficients (MEC). Due to these limitations, there has been an increased focus on metal-free organic sensitizers within the field. These organic sensitizers have several advantages, including low toxicity, structural flexibility, ecologically friendly features, high MECs, and simple synthesis techniques. In previous studies on the development of organic sensitizers, a range of chromophores like indoline, coumarin, phenothiazine, carbazole, triarylamine, and tetrathiafulvalene have been utilized [20,21].

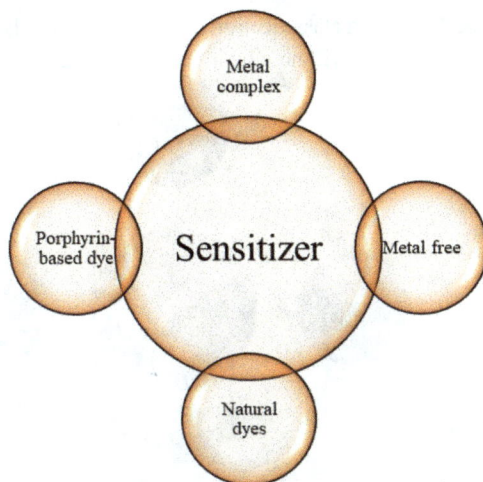

Figure 4. Types of sensitizers for DSSC

The clarification of structure-property relationships will play a crucial role in promoting the better utilization of these dyes in DSSCs. Undoubtedly, the future of DSSC technology is extremely promising, offering countless opportunities to enhance performance. It is essential to find a new way to design dyes without using metals, and this needs to take place right away. In recent scientific research, there has been a significant emphasis on studying carbazole, triphenylamine, and phenothiazine-based dyes. These works attempt to provide a complete understanding of this topic, catering to both the scientific and engineering communities. Furthermore, this research seeks to find prospective design elements that might be used in the production of novel dyes [22].

3.3 Electrolyte

Redox mediators play a crucial role in regenerating the excited dye molecules, and they are in the form of either liquid, gel, or solid electrolyte. On the other hand, the hole transport material (HTM) serves as a mediator for holes in solid electrolytes. The redox shuttle is responsible for regenerating the dye through the oxidation process and being reduced by the counter electrode. This process effectively regenerates the dye. The redox shuttle has a significant impact on the DSSC's photovoltage, which is a crucial parameter for the device's performance. It controls the electrochemical potential at the counter electrode by facilitating the recombination of electrons in the titanium dioxide with their oxidized counterparts. Figure 5 depicts a classification of the many electrolytes designed specifically for DSSCs. These electrolytes will now be investigated in greater depth.

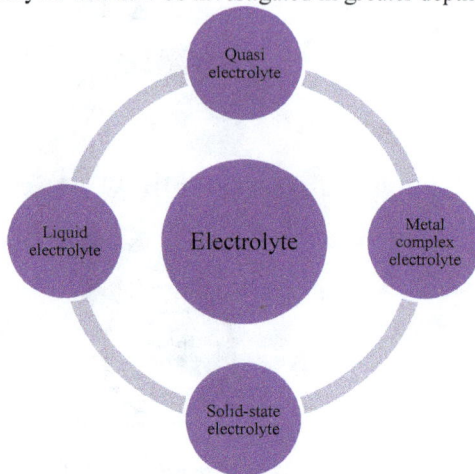

Figure 5. Types of electrolytes developed for efficient DSSC

Since the discovery of the DSSC, a number of redox shuttles have been utilized and investigated in DSSC electrolytes. Redox shuttles can be of the following types: i) Halide and pseudohalogens redox couple (e.g. I_3^-/I^-, Br_3^-/Br^- and $(SCN)_3^-/SCN^-$; ii) transition metal redox couples (e.g. Co^{3+}/Co^{2+}); iii) organic redox couples (e.g. TEMPO$^+$/TEMPO). These redox shuttles have been extensively studied in DSSCs [23]. These redox couples' potentials are listed in the Table 1.

Table. 1 Shows the redox potential of standard electrolyte Vs NHE

Electrolyte	Standard redox potential Vs NHE
I_3^-/I^-	~ 0.35 V
Br_3^-/Br^-	~ 1.1 V
Fe^{3+}/Fe^{2+} (complex)	1.37 V
Co^{3+}/Co^{2+} (complex)	~ 0.86
Cu^{2+}/Cu^+ (complex)	~ 0.97 V
(Fc^+/Fc)	~ 0.62 V
TEMPO$^+$/TEMPO	~ 0.8 V

3.4 Counter electrode

Counter electrode functions as a catalyst by transferring electrons from the external circuit to redox shuttle and completes the cycles. In solid-state DSSCs, this reduction occurs through an ionic transport medium, where the counter electrode obtains electrons from the oxidized redox pair. Secondly, it acts as a conduit for electrons, transporting them from the outer circuit to the electrolyte solution within the cell. Lastly, the counter electrode serves as a reflective surface, maximizing the utilization of solar energy. It reflects any unabsorbed light back, preventing its loss and enhancing the efficiency of the DSSC system. The counter electrode needs to possess several important characteristics in order to fulfil its basic functions. These include a high level of catalytic activity, conductivity, and reflectivity. Additionally, it should have a large surface area with a porous structure, as well as the ability to resist chemical corrosion and maintain high chemical and mechanical stability [24]. Moreover, the energy level of the counter electrode should align with the potential of the redox electrolyte. Lastly, it is crucial for the counter electrode to have good adhesion to the TCO substrate to effectively perform these essential functions. Properties of counter electrode summarizes in Figure 6.

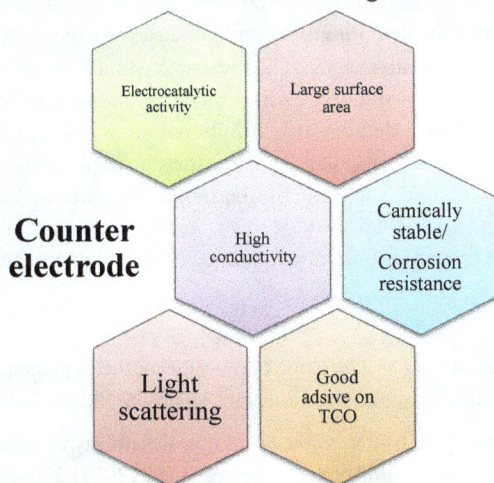

Figure 6. Properties of counter electrode for efficient DSSC

The adoption of platinum counter electrodes as the standard material for DSSCs has become widespread. Nevertheless, researchers are faced with a significant challenge in terms of the high cost and limited availability of platinum. This necessitates a need for future research on platinum counter electrodes to focus on the development of new techniques and composite materials that can reduce the amount of platinum used. Fortunately, several platinum-free catalysts can be employed as counter electrodes in DSSCs.

In recent research, there has been a strong emphasis on exploring alternative materials for counter electrodes in solar cells that do not require the use of platinum (Pt). These studies have primarily focused on various classes of transition metal compounds such as chalcogenides, phosphides, and oxides [25–27]. Additionally, binary or ternary metal oxides and sulphides have also been investigated as potential candidates. These materials are frequently mixed with different kinds of carbon allotropes to boost their efficiency in solar cell applications.

Conclusion and future perspective

In the coming years, dye-sensitized solar cells have the potential to significantly contribute to the production of sustainable energy. For the past thirty years, scientists and experts have been consistently exploring the possibilities of photovoltaic technology. In order to advance the field, it is imperative that future research focuses on the development of innovative ideas, techniques, and substances. This research should prioritize meeting various important criteria such as conductivity, catalytic activity, stability, efficiency, cost-effectiveness, and environmental cleanliness. Additionally, it should delve into understanding the intricate mechanisms that govern the creation, progression, and movement of photoinduced charge carriers. Furthermore, the interactions between photoanodes, electrolytes, and counter electrodes must be thoroughly examined, taking into consideration the unique properties and functionalities of each component.

Reference

[1] D. Oteng, Towards a sustainable PV waste policy: Exploring the management practices of end-of-life solar photovoltaic modules in Australia, PhD Thesis, 2023.

[2] M.A. Fazal, S. Rubaiee, Progress of PV cell technology: Feasibility of building materials, cost, performance, and stability, Solar Energy. 258 (2023) 203-219. https://doi.org/10.1016/j.solener.2023.04.066

[3] J. Penman, Performance evaluation of the photovoltaic system, (2023).

[4] F. Rehman, I.H. Syed, S. Khanam, S. Ijaz, H. Mehmood, M. Zubair, Y. Massoud, M.Q. Mehmood, Fourth Generation Solar Cells: A Review, Energy Advances. (2023). https://doi.org/10.1039/D3YA00179B

[5] A. Sharma, V.K. Bajpai, A Review on Development and Technology of Various Types of Solar PV Cell, Optimization Methods for Engineering Problems. (2023) 215. https://doi.org/10.1201/9781003300731-15

[6] B. O'regan, M. Grätzel, A low-cost, high-efficiency solar cell based on dye-sensitized colloidal TiO2 films, Nature. 353 (1991) 737-740. https://doi.org/10.1038/353737a0

[7] J. Gong, J. Liang, K. Sumathy, Review on dye-sensitized solar cells (DSSCs): Fundamental concepts and novel materials, Renewable and Sustainable Energy Reviews. 16 (2012) 5848-5860. https://doi.org/10.1016/j.rser.2012.04.044

[8] J.B. Asbury, E. Hao, Y. Wang, H.N. Ghosh, T. Lian, Ultrafast electron transfer dynamics from molecular adsorbates to semiconductor nanocrystalline thin films, The Journal of Physical Chemistry B. 105 (2001) 4545-4557. https://doi.org/10.1021/jp003485m

[9] V. Sugathan, E. John, K. Sudhakar, Recent improvements in dye sensitized solar cells: A review, Renewable and Sustainable Energy Reviews. 52 (2015) 54-64. https://doi.org/10.1016/j.rser.2015.07.076

[10] J.V. Vaghasiya, K.K. Sonigara, S.S. Soni, Role of metal oxides as photoelectrodes in dye-sensitized solar cells, in: Advances in Metal Oxides and Their Composites for Emerging Applications, Elsevier, 2022: pp. 287-338. https://doi.org/10.1016/B978-0-323-85705-5.00009-9

[11] V. Galstyan, J.M. Macak, T. Djenizian, Anodic TiO2 nanotubes: A promising material for energy conversion and storage, Applied Materials Today. 29 (2022) 101613. https://doi.org/10.1016/j.apmt.2022.101613

[12] A. Gopalraman, J.A. Raj, S. Karuppuchamy, S. Vijayaraghavan, Investigation on the effect of neodymium ion doping in TiO2 on the photovoltaic performance of dye-sensitized solar cells, Materials Chemistry and Physics. 292 (2022) 126785. https://doi.org/10.1016/j.matchemphys.2022.126785

[13] N. Kaur, D.P. Singh, A. Mahajan, Plasmonic engineering of TiO2 photoanodes for dye-sensitized solar cells: a review, Journal of Electronic Materials. 51 (2022) 4188-4206. https://doi.org/10.1007/s11664-022-09707-3

[14] R. Agarwal, S. Sahoo, V.R. Chitturi, J.D. Williams, O. Resto, R.S. Katiyar, Enhanced photovoltaic properties in graphitic carbon nanospheres networked TiO2 nanocomposite based dye sensitized solar cell, Journal of Alloys and Compounds. 641 (2015) 99-105. https://doi.org/10.1016/j.jallcom.2015.03.175

[15] M. Sufyan, U. Mehmood, S. Yasmeen, Y.Q. Gill, M. Sadiq, M. Ali, Metal-Oxide Semiconductor Nanomaterials as Alternative to Carbon Allotropes for Third-Generation Thin-Film Dye-Sensitized Solar Cells, in: Defect Engineering of Carbon Nanostructures, Springer, 2022: pp. 235-268. https://doi.org/10.1007/978-3-030-94375-2_9

[16] K.T. Dembele, G.S. Selopal, C. Soldano, R. Nechache, J.C. Rimada, I. Concina, G. Sberveglieri, F. Rosei, A. Vomiero, Hybrid carbon nanotubes-TiO2 photoanodes for high efficiency dye-sensitized solar cells, The Journal of Physical Chemistry C. 117 (2013) 14510-14517. https://doi.org/10.1021/jp403553t

[17] G. Speranza, W. Liu, L. Minati, Applications of Plasma Technologies to Material Processing, CRC Press, 2019. https://doi.org/10.1201/9780429264658

[18] B.T. Huy, D.H. Kwon, S.-S. Lee, V.-D. Dao, H.B. Truong, Y.-I. Lee, Optical properties of Sr2YF7 material doped with Yb3+, Er3+, and Eu3+ ions for solar cell application, Journal of Alloys and Compounds. 897 (2022) 163189. https://doi.org/10.1016/j.jallcom.2021.163189

[19] N. POOTRAkULCHOTE, Investigation on Functionalized Ruthenium-Based Sensitizers to Enhance Performance and Robustness of Dye-Sensitized Solar Cells, EPFL, 2012.

[20] G. Sharma, V. Singh, S.N. Dolia, I.P. Jain, P.K. Jain, C. Lal, Present status of metal-free photosensitizers for dye-sensitized solar cells, Materials Today: Proceedings. (2023). https://doi.org/10.1016/j.matpr.2023.02.179

[21] A. Grobelny, Z. Shen, F.T. Eickemeyer, N.F. Antariksa, S. Zapotoczny, S.M. Zakeeruddin, M. Grätzel, A Molecularly Tailored Photosensitizer with an Efficiency of 13.2% for Dye-Sensitized Solar Cells, Advanced Materials. 35 (2023) 2207785. https://doi.org/10.1002/adma.202207785

[22] M. Yahya, A. Bouziani, C. Ocak, Z. Seferoğlu, M. Sillanpää, Organic/metal-organic photosensitizers for dye-sensitized solar cells (DSSC): Recent developments, new trends, and future perceptions, Dyes and Pigments. 192 (2021) 109227. https://doi.org/10.1016/j.dyepig.2021.109227

[23] Masud, H.K. Kim, Redox Shuttle-Based Electrolytes for Dye-Sensitized Solar Cells: Comprehensive Guidance, Recent Progress, and Future Perspective, ACS Omega. 8 (2023) 6139-6163. https://doi.org/10.1021/acsomega.2c06843

[24] M. Kokkonen, P. Talebi, J. Zhou, S. Asgari, S.A. Soomro, F. Elsehrawy, J. Halme, S. Ahmad, A. Hagfeldt, S.G. Hashmi, Advanced research trends in dye-sensitized solar

cells, Journal of Materials Chemistry A. 9 (2021) 10527-10545.
https://doi.org/10.1039/D1TA00690H

[25] A.R. Tapa, W. Xiang, X. Zhao, Metal Chalcogenides (M x E y; E= S, Se, and Te) as
Counter Electrodes for Dye-Sensitized Solar Cells: An Overview and Guidelines,
Advanced Energy and Sustainability Research. 2 (2021) 2100056.
https://doi.org/10.1002/aesr.202100056

[26] X. Wang, B. Zhao, W. Kan, Y. Xie, K. Pan, Review on low-cost counter electrode
materials for dye-sensitized solar cells: effective strategy to improve photovoltaic
performance, Advanced Materials Interfaces. 9 (2022) 2101229.
https://doi.org/10.1002/admi.202101229

[27] A. Yildiz, T. Chouki, A. Atli, M. Harb, S.W. Verbruggen, R. Ninakanti, S. Emin,
Efficient iron phosphide catalyst as a counter electrode in dye-sensitized solar cells,
ACS Applied Energy Materials. 4 (2021) 10618-10626.
https://doi.org/10.1021/acsaem.1c01628

Materials Research Forum LLC
https://doi.org/10.21741/9781644903032-3

Chapter 3

Polymer/Organic Solar Cells: Progress and Current Status

L. Vidhya [1*], S. Vinodha[2], S.J. Pradeeba[3], R. Jeba Beula[4]

[1, 3]Department of Science and Humanities, Hindusthan College of Engineering and Technology, Pollachi Main Rd, Coimbatore-641050, Tamil Nadu, India

[2]Department of Science and Humanities, Jayaraj Annapackiam CSI College of Engineering, Nazareth, Thoothukudi, Tamil Nadu, India

[4]Department of Physics, Karunya Institute of Technology, Coimbatore, Tamil Nadu, India

vidhuram236@gmail.com , vinodha.harris@gmail.com,

Abstract

Organic solar cells (OSCs) have sparked widespread interest in recent decades due to benefits such as low cost, flexibility, semitransparency, non-toxicity, and suitability for roll-to-roll large-scale production. The development of OSCs with high-performance active layer materials, electrodes, and interlayers as well as innovative device architectures has made significant strides. In particular, the advancement of OSCs' power conversion efficiency (PCE) has been greatly aided by the development of active layer materials, including novel acceptors and donors. Photovoltaic cells are one of the most promising renewable energy sources for resolving energy and environmental challenges. Organic solar cells (OSCs) have several advantages over other photovoltaic technologies, including low cost, lightweight, semi-transparency, and flexibility. This final benefit, which results from the inherent flexibility of organic active layers, is unique to OSCs. Flexible OSCs (F-OSCs), which have intriguing applications in areas like wearable electronics and building-integrated photovoltaic, have progressed quickly in recent years, and great progress has been made in this area. In this chapter, we provide an overview of current developments in semi-transparent organic solar cells, polymer-based solar cells and their Fullerene-containing polymers for organic solar cell. Additionally, a brief discussion of fullerene-containing polymers for organic solar cell was specified. The final step is the presentation of difficulties for the advancement of F-OSCs.

Third Generation Photovoltaic Technology   Materials Research Forum LLC
Materials Research Foundations 163 (2024) 52-91   https://doi.org/10.21741/9781644903032-3

Keywords

Organic Solar Cells, Photovoltaic Cells, Polymer-based Solar Cells, Fullerene-containing Polymers

Contents

1. Introduction

In the past two decades, organic solar cells (OSCs) have developed quickly, and the power conversion efficiency (PCE) of single-junction OSCs has lately acquired close to 20% efficiency. Two important components of this progression are the use of innovative materials and device engineering. This chapter provides a systematic summary of device engineering, encompassing morphological characterization and optimization, flexible and large-area OSCs, device physics, and OSC stability. Solar cell efficiencies, in accordance with Abodunrin et al., [1] are currently significantly restricting the commercialization of these devices. The search for a superior energy source has led to the development of numerous different technologies. Some of these technologies have been shown to be hazardous to life. For instance, the use of fossil fuel generators is more common in developing countries. As a result of this technology, the global carbon footprint has significantly expanded. Nuclear energy technologies are still used in the majority of nations worldwide [2]. The main problem with nuclear energy is the radioactive waste, which is toxic. Radioactive waste is any material, whether it be solid, liquid, or gaseous, that contains a radioactive nuclear substance. The improper management of this trash could lead to pollution of the air, land, and water. An interdisciplinary field made up of materials science, chemistry, physics, engineering, etc. includes organic solar cells (OSCs). In a nutshell, material science and device engineering make up the research area of OSCs. A comprehensive overview of the material science aspect of OSCs, including tiny molecule donors and acceptors, conjugated polymer donors and acceptors, and interface materials, may be found in [3]. The performance of the device depends critically on the active layer's nano-morphology. Typically, morphology optimization through thermal annealing, additive treatment and the like can be used to achieve an optimum continuous interpenetrating network nano-morphology of the active layer, which can significantly alter the phase separation and promote charge transfer and collection, among other things. Additionally, atomic force microscopy and optical microscopy (OM) can be used to characterize the development of the nano-morphology of the active layer. Scaling-up of OSCs and flexible OSCs have attracted a lot of attention recently for further practical applications. Furthermore, OSC stability and degradation are progressively gaining importance. Due to losses in electrical, geometric, optical, and other ways, including flexible and big area OSCs often have lower PCEs than small area ones.

Recently, intense research has focused on constructing a variety of organic electronics components, such as light-emitting diodes, field-effect transistors, memory cells, sensors, and solar cells, using organic materials with semiconductor capabilities. Conjugated polymers are being studied as an alternative to organic semiconductors following the introduction of conductive polyacetylene [4]. Since the mid-1990s, when conductive conjugate was created, meaningful progress in the creation of organic solar cells (OSCs) has been conceivable. The benefits listed above make it possible to produce an OSC utilizing relatively inexpensive techniques such inkjet printing and stamping technology, in addition to the ability to deposit films from solutions at normal pressure onto flexible substrates of a large area [5].

Due to their potential in manufacturing large-area, low-cost solar cells, organic photovoltaic devices have recently attracted a lot of attention. Their power conversion efficiencies have significantly improved since the first studies on molecular thin film devices more than 30 years ago, rising from 0.001% in 1975 [6] to 1% in 1986 [7] and more recently to 5.5% in 2006 [8-11]. They may soon be a competitive alternative to inorganic solar cells due to improvements in efficiency. Small molecules [12-15, 16], conjugated polymers [17], small molecules and conjugated polymers combined [18-20], or combinations of organic and inorganic materials [21] have all been used as the active layer in various published concepts. The layer in which the bulk of incident light is absorbed and charges are produced is referred to as the "active layer" in this context. The molecular weights of polymers and small molecules are different. More often than not, macromolecules with molecular weight greater than 10,000 amu are referred to as polymers, while smaller molecules are referred to as "oligomers" or "small molecules."

Organic materials having semiconductor capabilities have become the recent focus of intense research aimed at developing various components of organic electronics, such as field-effect transistors, memory cells, light-emitting diodes, sensors and solar cells. Conjugated polymers have been investigated as a replacement for organic silicon semiconductors after the introduction of conductive polyacetylene [22]. Since the mid-1990s, when conductive conjugate was created, meaningful progress in the creation of organic solar cells (OSCs) has been conceivable. The benefits listed above make it possible to produce an OSC utilizing relatively inexpensive techniques such inkjet printing and stamping technology, along with the ability to deposit films from solutions at normal pressure onto flexible substrates of a large area [23].

2.   Progress on organic solar cells

The three types of organic solar cells include single, bilayer, and bulk heterojunction organic solar cells. Single-layer OSC is the simplest organic solar cell structure [24]. A single layer of organic material forms the basis of the most basic OSC structure. A conjugated polymer film is sandwiched between two conducting electrodes to create the device. Figure 1(A) shows a single layer device's basic configuration. It is made up of an organic absorber medium surrounded by aluminum and indium tin oxide (ITO) electrodes. This device's primary drawback is the absorber medium's thin thickness, which ranges from 10 to 20 nm. Two distinct organic layers are sandwiched between the electron and hole collecting layers in the bilayer OSC. The device's fundamental construction is shown in Figure 1 (B). At the point where the donor and acceptor interface, excitation separation takes place. The electron and hole collecting layers change the electrodes' work function to produce improved ohmic connections. The acceptors display higher ionization potentials and electron affinities than the donors. When a photon strikes the donor layer, it is absorbed, and the donor material's electrons are stirred up from the highest occupied molecular orbital (HOMO) level to the lowest unoccupied molecular orbital (LUMO) level, creating exciton. Last but not least, a polymer blend serves as the photoactive (absorber) layer in bulk heterojunction OSC. Electron acceptor and donor polymers are often combined to create the mix, which is then placed between the electrodes [25].

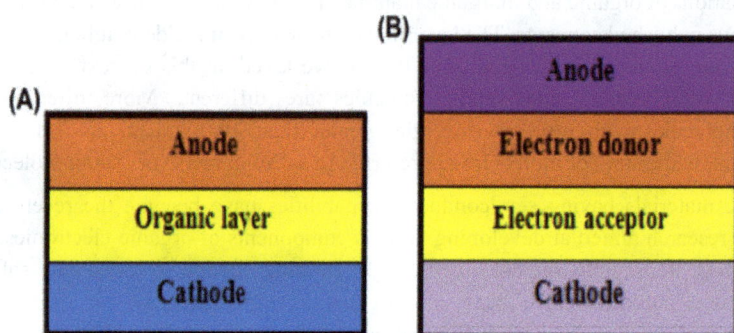

Figure 1 : Different types of organic solar cells: (A) single layer and (B) bilayer (Reproduced from Progress on Organic Solar Cells: A Short Review)(Open Access) [26].

3.   Polymer-based solar cells

Polymers have typically been regarded as electrical insulators throughout their existence. An intense study on conjugated polymers began in 1977 when the team of Heeger, Shirakawa, and MacDiarmid found that doped polyacetylene could reach metallic conductivity [27]. Later, in 2000, the effort of these three pioneers was recognized with the Nobel Prize (Chemistry). From this earlier study, very pure and soluble conjugated polymers did not become generally accessible until the late 1990s. While the early focus of the research was mostly on chemically doping conjugated polymers to increase their conductivity, innately semiconducting and highly soluble polymers also attracted significant attention. Conjugated polymers have since been used in a variety of semiconductor devices; including thin film transistors (TFTs), light-emitting diodes (LEDs), and solar cells were all thoroughly examined. These polymers' conjugated structures often have chains and rings of carbon atoms hybridized in sp1 and sp2, resulting in a framework of alternate double- or triple- and single-bonds. In this illustration, single bonds are shown as σ-bonds, and double bonds are made up of a σ-bond and a π-bond. The π-electrons, however, are delocalized across the entire molecule and are not restricted to the specific bonds. As a result, conjugated polymers can be compared to semiconductors that are almost one-dimensional. The band gap is determined by the bond alternation and the degree of π-orbital overlap. By spin-coating from a mixture of an electron-accepting polymer and an electron-donating polymer, layers containing so-called polymer blends can be created. Solid polymer blends frequently phase-separate due to the entropy of mixing. This results in a distributed bulk heterojunction owing to the low entropy of polymers. Without any extra processing steps, a huge acceptor-donor interface is generated. Nearly all exciton can be activated if the acceptor and donor components are thoroughly mixed, and the length scale of the phase separation is within the range of the exciton diffusion length (10 nm). Furthermore, the morphology formed when a thin layer of immiscible polymers is deposited from the solution depends greatly on a number of factors, including, the interaction of the polymer with the substrate surface, the solubility of the individual polymers in the solvent, the deposition technique, the layer thickness, and the drying procedure.

Table 1: Best-in-class solar cells: small molecule-based solar cells. (Reproduced from Recent Advances in Organic Solar Cell, Hindawi Publication, open access journal) [28].

Donor	Acceptor	η	V_{oc}	FF	IPCE
CuPc	C60	5.7%	1.0 V	59%	NA
CuPc	C60	5.0%	0.6 V	60%	64%
MeO-TPD, ZnPc (stacked)	C60	3.8%	1.0 V	47%	NA
CuPc	C60	3.5%	0.5 V	46%	NA
DCV5T	C60	3.4%	1.0 V	49%	52%
CuPc	PTCBI	2.7%	0.5 V	58%	NA
SubPc	C60	2.1%	1.0 V	57%	NA
MeO-TPD, ZnPc	C60	2.1%	0.5 V	37%	NA
TDCV-TPA	C60	1.9%	1.2 V	28%	NA
Pentacene on PET	C60	1.6%	0.3 V	48%	30%
SnPc	C60	1.0%	0.4 V	50%	21%

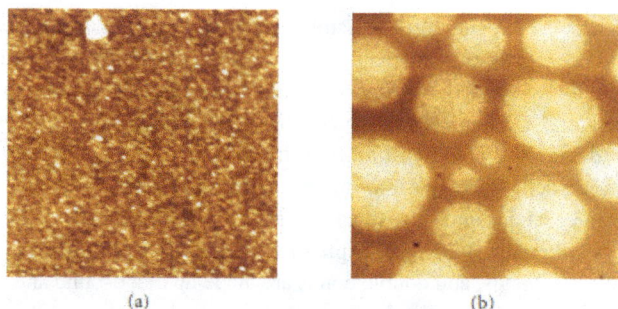

(a)   (b)

Figure 2: The AFM height images (size 2.8 × 2.8 μm) reveal different phase-separated morphologies for blends of PFB and F8BT (1:1 ratio by mass) spin-coated from (a) chloroform and (b) xylene (Reproduced from Recent Advances in Organic Solar Cell, Hindawi Publication, open access journal) [28].

Therefore, it is frequently arbitrary and, depending on trial and error, to modify the length scale of phase separation in thin layers. The potential for remixing is caused by the polymers' macromolecular structure. The Polaris ability or differing polarity of the polymer repeat units, which makes contact between similar polymers energetically advantageous, is one factor contributing to phase separation. However, many other molecules can combine in any proportion, including water and ethanol. For these small molecules, the enthalpic interaction is defeated by the rise in entropy that results from mixing. The Flory and Huggins theory can be used to explain a pair of polymers' mixing characteristics. However, the true scenario for thin blend films is significantly more complicated since the interactions of the two polymers with the substrate and the surface, in addition to the solvent influence and the solvent evaporation rate, must be considered. Therefore, it is challenging to forecast or even control the morphology due to the interaction between mixing and dewetting. The combination of PFB and F8BT [poly(9,9-dioctylfluorene-2,7-diyl-co-benzothiadiazole)] is the most thoroughly researched conjugated polymer system to date. The length scale on which phase separation occurs strongly depends on interaction with the substrate, the solvent utilized and deposition conditions and according to investigations on its blend morphology in relation to the solar cell attributes. While the length scale of the phase separation for films created from chloroform is less than 100 nm. Spin-coating a 1:1 PFB:F8BT solution in xylene results in phase separation at the micrometer scale, as seen in the atomic force microscopy (AFM) image in Figure 2. This phenomenon resulted from the low boiling point solvent chloroform evaporating more quickly than xylene, which stopped the diffusion and large-scale restructuring of the polymer chains and caused a finer scale of the phase separation. The start of spinodal composition is prevented when the substrate is heated during spincoating because the faster evaporation causes a phase separation on an even smaller scale. It is interesting to note that the PFB:F8BT blend layer spin-coated from xylene phase-separated morphology exhibited a substructure at the nanometer scale in addition to micrometer sized features. It was suggested that one phase contains roughly equal amounts of PFB and F8BT based on AFM and Raman microscope research, but the second is F8BT rich and contains 80% F8BT and just 20% PFB. It was also discovered that this second phase had a greater thickness. The blend layers can be defined as PFB-rich cylinders encircled by a F8BT-rich matrix, according to further study. As anticipated, preparation circumstances had a significant impact on the solar cell's characteristics. Compared to xylene solutions (IPCE = 0.5-1%) the IPCE was roughly twice as high for layers spin-coated from chloroform solutions (IPCE = 2-4.5%) for the PFB:F8BT (1:1) blends. This was explained by a greater phase separation for the xylene-prepared layers. In later research, photovoltaic devices were examined using various PFB and F8BT ratios in layers made from xylene solution. It was discovered that the layer composition had a significant impact on both the phase separation and the external quantum efficiencies. The devices with the highest efficiency were those with five times as much of the electron-transporting F8BT as PFB. The photocurrent is not boosted at the interface of the two domains, according to a more recent study that used near field photoconductivity measurements to investigate the photocurrent generation. Instead, it was suggested that although photocurrent appears to be produced in both phases, the phase with the smaller height appears to be more effective. These investigations conclusively demonstrated that the preparation circumstances and layer composition have a significant and, in part, unanticipated impact on the device's photovoltaic capabilities. A study for a fixed length scale of phase separation but different blend ratios was also not attainable by utilizing different layers spin-coated from organic solvents because it has been demonstrated that the mesoscale phase separation substantially depends on the blend ratio. Therefore, a technique for precise control of the phase separation that develops in polymer blends would be greatly desired.

3.1 Polymer solar cell devices

The electron acceptor was CN-PPV (cyano-para-phenylenevinylene), while the electron donor was MEH-PPV (Poly(2-methoxy-5-(2'-ethyl-hexyloxy)-1,4-phenylene-vinylene). External quantum efficiencies were still poor (about 5-6%), which was most likely caused by an unoptimized nanomorphology. After the CN-PPV was altered by the addition of an ether group to boost solubility, a significant increase in efficiency was attained. Breeze *et al.* exhibited external quantum efficiencies of 24% in 2000 using the copolymer M3EH-PPV as the donor and CN-Ether-PPV as the acceptor (Figure 3 for the chemical structures). The same authors published even higher efficiency in 2004. In that research, it was shown

Materials Research Forum LLC
https://doi.org/10.21741/9781644903032-3

for the first time that a polymer-polymer blend device was capable of achieving 1% power conversion efficiency. Although the open circuit voltage was 1 V, the fill factor was only 25%, which left an opportunity for improvement. By combining the same materials with enhanced processing, Kietzke et al. [29] achieved the best power conversion efficiency to date for solar cells made of polymer blends in 2005. It could be demonstrated that annealing the layers results in a two-fold increase in efficiency over the prepared layers. White light conversion efficiency was 1.7% with an open circuit voltage of 1.36 V. The fill factor increased to 35 %, which indicates better charge transmission. According to photophysical experiments on PPV-based blends, the exciplex production in this system may have been a significant contributor to loss. Recently, Koetse *et al.* [30] reported solar cells based on a new acceptor copolymer called PF1CVTP with MDMO-PPV as the donor. A significant 42% quantum efficiency was attained. The power conversion efficiency, however, was unable to surpass previous records since it started to decline at greater light intensities typical of solar illumination. Polymer bilayer structures have also been studied using a variety of methods. According to recent investigations, devices using BBL as the electron acceptor and insoluble PPV as the electron donor, methane sulfonic acid provided the deposit for BBL.

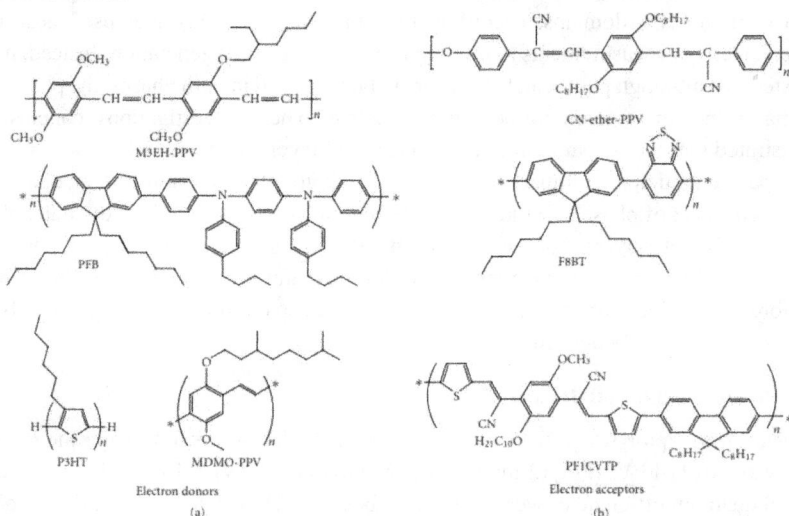

Figure 3: Chemical structures of different conjugated polymers employed in organic solar cells (Reproduced from Recent Advances in Organic Solar Cell, Hindawi Publication, open access journal) [28].

Extremely high quantum efficiencies of up to 62% were demonstrated by the devices. Unfortunately, with increasing light intensities, the device's efficiency substantially dropped. Under a typical 1 sun, the power conversion efficiency fell from a record-breaking 5% at very low light intensities to 1.5%. Similar efficiency levels could be attained using CN-Ether-PPV as the acceptor and M3EH-PPV as the donor. There is a lot of space for improvement with polymer solar cells (the best of their kind's defining characteristics are shown in Tables 2 and 3). To achieve 6-7% power conversion efficiency, the fill factor and quantum efficiency must both be increased. Recent investigations revealed that the low dissociation efficiency of the photo-generated exciton into free charge carriers is now the main factor limiting the device's performance. Due to the amorphous structure of the electron-accepting polymers, even after dissociation, the electrons have a tendency to concentrate close to the heterointerface. More crystalline electron acceptor polymers with better electron mobilities are required to advance to higher efficiencies.

Table 2: Best in class solar cells: polymer-polymer (blend) solar cells. (Reproduced from Recent Advances in Organic Solar Cell, Hindawi Publication, open access journal) [28].

Donor	Acceptor	η	V_{oc}	FF	IPCE
M3EH-PPV	CN-Ether-PPV	1.7%	1.4 V	35%	31%
MDMO-PPV	PF1CVTP	1.5%	1.4 V	37%	42%
M3EH-PPV	CN-Ether-PPV	1.0%	1.0 V	25%	24%

Table 3: Best in class solar cells: polymer-polymer (bilayer) solar cells. (Reproduced from Recent Advances in Organic Solar Cell, Hindawi Publication, open access journal) [28].

Donor	Acceptor	η	V_{oc}	FF	IPCE
PPV	BBL	1.5%	1.1 V	50%	62%
MDMO-PPV:PF1CVTP	PF1CVTP	1.4%	1.4 V	34%	52%
M3EH-PPV	CN-Ether-PPV	1.3%	1.3 V	31%	29%
MEH-PPV	BBL	1.1%	0.9 V	47%	52%
M3EH-PPV	CN-PPV-PPE	0.6%	1.5 V	23%	23%

4.   Stability of the OSC

Due to PCE's rapid progress, OSC stability has superseded commercialization as the most crucial problem. OSCs are thin, organic or inorganic films with a diameter of a few nanometers. These nano-thin films are vulnerable to performance degradation under a variety of stress conditions, such as thermal heating, water, light, oxygen, and even electric field. Through a number of paths, this can lead to a slow or quick performance decline. The effects of oxygen and water, which can damage cells irreversibly through chemical oxidation of various components, are made severe when light irradiation is introduced. They interact with metal electrodes, corroding them; change the electrical characteristics of an interfacial PEDOT: PSS layer by absorbing water; or oxidize conjugated organic semiconductors, causing the breakdown of organic materials. Good encapsulation can separate the cell from the oxygen/water. This will stop the degradation of the device by blocking the exposure to oxygen/water. The literature demonstrates numerous encapsulating methods. For instance, Tsai and Chang [31] reported on the use of atomic layer deposition (ALD) to produce composite Al2O3/HfO2 encapsulating layers for P3HT: PC61BM cells. The 26 nm Al_2O_3/HfO_2 nano-composite encapsulating film showed a water vapour transmission rate (WVTR) (5 104 g/(m2 day)), successfully preventing the aggression of O_2/H_2O after being stored for more than 10,000 hours in ambient air under conditions that speed up aging or for more than 1,800 hours in a 65 °C/60% RH. According to the findings, flexible P3HT: PC61BM cells sandwiched between two barrier films preserved >95% of the device's initial performance after 1,000 hours of aging under damp-heat (85 °C/85% RH) conditions. Recently, a solution-processed multilayered barrier film made from perhydropolysilazane (PHPS) ink was converted into a silica layer by intense UV irradiation. In moist heat circumstances, P3HT: PC61BM cells showed enhanced device lifetime for more than 300 hours with these solution-processed barrier layers. While operating, polymer solar cells also come into contact with the stressors like heat, light, and electric field, which inevitably cause performance to degrade. For instance, Guldi et al. [32] demonstrated that photon-induced dimerization of PC61BM occurred in fullerene-based cells, resulting in a reduction in charge carrier mobility and an alteration in the fill factor and short circuit current of OSCs. Thermally-induced morphological change in the photoactive layer was determined to be the reason why both fullerene and non-fullerene solar cells' performance declined. The performance decay of polymer solar cells is external load-dependent, which suggests the electric fields strength will also be a factor, according to Yan et al.'s research [33]. In contrast to water and oxygen, light (to create charges inside the photoactive layer), heat (temperature rises under light irradiation), and electric fields (there are always internal and external) are invariably present when a cell is working. Additionally, encapsulation does not avoid these stress factors. Therefore, it is believed that the degradation of PSC performance caused by light, heat, and electric fields is an inherent or intrinsic decay process that is closely related to the stability of the materials used and the layered structure of the devices. The anticipated lifetime of the thermally evaporated organic solar cells is over 27,000 years, which suggests that they may be sufficiently intrinsically stable for usage. However, the majority of solution-processed polymer solar cells did show a T80 lifetime, which is the time in which the cell performs at 80% of its peak capacity and is often fewer than 1,000 h. Therefore, it is still essential to understand the PSCs' inherent deterioration processes and find suitable countermeasures.

4.1   Air stability of the OSCs

Under ambient conditions, organic materials can react with oxygen, and oxidation reactions can lower OSC performance. Particularly because traditional photo-active layers, such PC61BM and PC71BM, are made of fullerene derivatives, self-aggregation of these materials is possible when they are exposed to oxygen. Therefore, it is crucial to develop acceptors with novel chemical structures in order to boost the intrinsic stability of photo-active materials (i.e., acceptors) for industrial applications. Massive efforts have been made to create robust NFAs that can support the long-term operation of high-performance OSCs. The O-IDTBR NFA was created by Holliday et al. [34] who also compared the air stability of O-IDTBR-equipped devices to those with fullerene acceptors in an inverted configuration. The PCE of the fullerene derivative-based devices gradually declined, and after 1200 hours, the devices were scarcely functional. The device based on O-IDTBR, however, worked steadily for 1200 hours while preserving 72% of the original PCE, with the exception of a tiny reduction at the beginning. This was attributed to the active layer's flat surface because O-IDTBR does not aggregate in ambient settings. According to earlier findings, utilizing a freshly synthesized donor material, it was possible to achieve long-term stability of a device using EH-IDTBR. The stability of highly efficient OSCs with PCEs over 10% was assessed while various NFA structures were reported. Only 10% of PCE was reduced in OSCs based on PTB7-Th:COi8DFIC or PM6:BP-4F after 30 days of air storage. Excellent storage durations of 30 days were demonstrated by Cai et al. [35] for PBDB-T:BTP-4F (with INB-1F, -3F, and 5F additives) with starting PCEs greater than 15%. Durability tests of the OSCs at ambient conditions without any encapsulation were done to further illustrate the inherent stability of NFAs. By optimizing the inherent stability of the NFA materials, stable functioning of the F13:Y6 BHJ and L2:TTPT-T-4F BHJ systems with a high PCE above 13% was observed over a period of 1000 hours without encapsulation. A significant barrier to the commercialization of OSCs is air stability, which has mostly been removed by changing the inherent characteristics of the NFA materials. Beyond OSCs, this is anticipated to usher in a new era of organic electronics made up of different organic semi-conductors.

4.2   Thermal stability of the OSCs

Organic materials are susceptible to severe degradation due to thermal stresses. The photo-active material must, however, have great thermal stability because the OSCs generate a lot of heat during operation. Due to their structural flexibility, NFAs offer a solution to the thermal stability issue. Polymer NFAs were initially employed to increase thermal stability, but the device based on polymer NFAs could only sustain 70% of the initial PCE at 180°C for 20 hours. A novel NFA, SF(DPPB)4, was created for great thermal stability. When combined with P3HT, the SF(DPPB)4 device displayed a Voc of 1.14 V and PCE of 5.16%. Particularly, devices based on SF(DPPB)4 resisted a heat treatment of 150 °C for 3 h without PCE loss, while those based on fullerene derivatives showed a reduced PCE. Ladder-type donor units (like IDT) were developed to generate low-crystalline materials without aggregation, enhancing the thermal stability of NFAs. As a result, the device using the IDT-BT-R acceptor maintained a similar PCE of 8.2% at 150°C for a period of time. Other investigations added polymer additives to BHJ systems based on ITIC. Block copolymers, for instance, function as compatibilizers to stop the phases from aggregating inside the BHJ. The device with ITIC and compatibilizer maintains 77% performance following heat treatment at 100°C for 120 h in a N2 atmosphere, despite burn-in loss being initially noticed. Additionally, according to reports, it put forth the ITYM all-fused-ring electron acceptor (AFRA), a new type. Single-crystal X-ray research, which reveals a planar noncyclic structure with strong "-" stacking, supports it. When compared to the traditional carbon-bridged INCN-type acceptors, ITYM displays superior thermal and chemical stability with extremely promising performance.

4.3   Light stability of the OSCs

Continuous lighting is required throughout operation because solar cells need sunshine to convert solar energy into electrical energy. However, sunshine causes photoelectric effects and the deterioration of biological molecules. Therefore, it is critical to optimize the molecular structure in the commercialization of OSCs in order to reduce the harm brought on by illumination. ITIC, when combined with PTB7-Th among the several NFA units, caused burn-in loss and inadequate light stability after just one hour of illumination. Additional research on ITIC degradation using a comparison to FAs (Fig. 4 (a and b)). The blend film's photodegradation or aggregation with donor materials might be sped up by the ITIC materials. ITIC's photodegradation mechanism was studied by Park et al. [36], who found that interactions with hydroxyl radicals led to the breakdown of the double bond in ITIC.

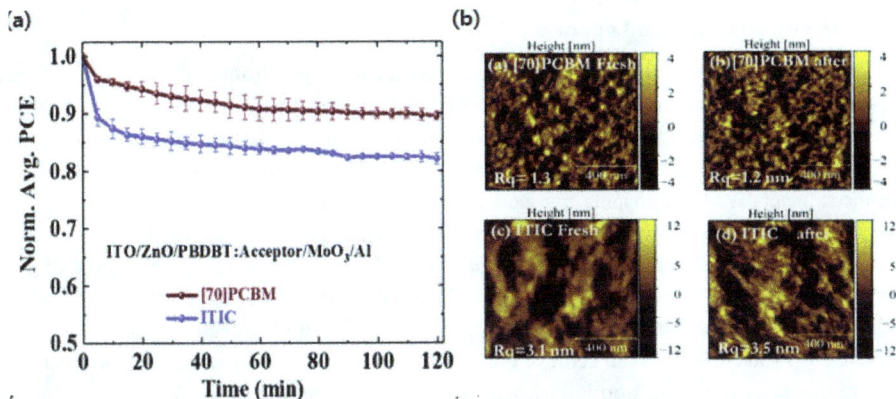

Figure 4. (a) Performance of PBDB-T-based inverted OSCs under continuous illumination. (b) AFM images of fresh and exposed organic thin films (PC70BM and ITIC). Reprinted with permission [37] Copyright © 2022 American Chemical Society

As a result, adding stable functional groups to NFAs is required to improve their photo-stability. Additionally, it demonstrated photo-stable devices made of different ITIC derivatives (ITIC, ITIC-2F, ITIC-M, ITIC-DM, and ITIC-Th) in a N2 environment with constant irradiation from one sun. Despite the fact that PBDB-T is comparatively photo-stable, the ITIC-DM-based device quickly lost performance as a result of its photo-catalytic activity. Other materials, particularly ITIC-2F, performed very steadily even after 1600 hours of illumination. It was established that the structural design of the NFAs has a significant impact on the photo-stability. Utilizing blend systems with PBDB-T and PBDB-T-2F donors, N2200 was used to improve the photo-stability of ITIC-M. The ITIC-M improved the devices' series resistance, and the N2200 served as a stabilizer to reduce photodegradation. The photo oxidation reaction sites and processes for ITIC and fullerene derivatives are depicted in earlier studies. Due to three-dimensional molecular packing with donor materials (mostly PTB7-Th), EH-IDTBR-based devices showed high illumination stability. By adding ring-locked C-C double bonds between D-A conjugation, a new molecular design strategy has been reported to increase the intrinsic chemical and photochemical stability of A-D-A type NFAs. This is attributed to increased steric hindrance of nucleophilic attack and the formation of intramolecular C-H/O interactions. The photo stability of the PTB7-Th:IDTT-CT based OSCs was very encouraging, with the PCE able to maintain >80% of the original values after 200 hours of direct sunlight in air without a UV filter. The success of the ring-locking design technique is suggested by the fact that this photo stability performance significantly outperforms that of standard NFAs like ITIC, IT-4F, and IT-M.

5. Device structure and operation

Many essential characteristics set organic semiconductors apart from traditional crystalline in-organic semiconductors (like silicon) [28]. First off, crystalline inorganic semiconductors have mobilities that are several orders of magnitude higher than those of organic semiconductors. The easiest method to describe the transport processes in organic semiconductors is to use hopping transport instead of band transport, which is present in most crystalline inorganic semiconductors. Even for organic semiconductors, the maximum recorded hole mobilities (h) are currently only about 15 $cm^2V^{-1}S^{-1}$ for small molecule single crystals and 0.6 $cm^2V^{-1}S^{-1}$ for liquid crystalline polymers (silicon: h = 450 $cm^2 V^{-1} S^{-1}$). Only specific TFT architectures using extremely crystalline tiny molecules (such as silicon, where e = 1400 $cm^2V^{-1}s^{-1}$) can achieve greater values for organic materials' highest electron mobilities (e), which are normally lower and hover around 0.1 $cm^2V^{-1}s^{-1}$. The mobility values for the most popularly used amorphous organic materials in organic solar cells are even significantly lower. The organic layer in solar cells can only be several hundred nanometers thick due to the poor mobilities. The good news is that organic semiconductors are potent UV-VIS absorbers. So, for optimal absorption, only organic layers about 100 nm thick are required. Second, compared to, say, silicon, organic semiconductors have substantially greater exciton binding energies. A hole is left in the highest occupied molecular orbital (HOMO) when an organic semiconductor absorbs a photon with enough energy to promote an electron into the lowest unoccupied molecular orbital (LUMO). However, this electron-hole pair develops a strongly bonded state as a result of electrostatic interactions state which is called singlet exciton. Although there is some controversy over the exact binding energy of this exciton, it is assumed to be in the 200–500 meV range. The exciton binding energy for organic semiconductors is roughly an order of magnitude higher than that for inorganic semiconductors like silicon, where light excitations frequently lead to free carriers at ambient temperature. The thermal energy at room temperature (25 meV) is inadequate to successfully induce free charge carriers in organic materials through exciton dissociation, even at average internal electric fields (106–107 V/m). According to studies using the frequently used polymer poly(2-methoxy-5-(2'-ethyl-hexyloxy)-p-phenylene vinylene (MEH-PPV), only 10% of the exciton dissociate into free carriers in a pure layer, with the remaining exciton decaying via radioactive or nonradioactive recombination pathways. Because of this, single-layer polymer device energy efficiencies are often sub 0.1%. The most significant finding on the path to high efficiency organic solar cells was that heterojunction solar cells, which combine organic materials that receive both holes and electrons, performed significantly better than single component devices. While most exciton in single component devices quickly recombine, using the heterojunction technique, photo-generated exciton (bound electron-hole pairs) in the polymer layer can be effectively separated into free carriers at the interface. The charge separation occurs between the donor and acceptor molecules through the action of a large potential drop. If the potential difference between the donor's ionization potential and the acceptor's electron affinity is greater than the exciton binding energy, an electron can jump from the LUMO of the donor (the material with the higher LUMO) to the LUMO of the acceptor after being photo excited from the HOMO to the LUMO (Figure 5). However, because of the donor's larger HOMO level, this procedure—known as photo induced charge transfer—can only result in free charges if the hole is still there. In contrast, the exciton totally transfers to the material of the lower-band gap with energy loss if the acceptor's HOMO is higher. For effective exciton dissociation at the heterojunction, the donor and acceptor materials need to be close to one another. Typically, the optimal length scale falls within the exciton diffusion length range, which is a few tens of nanometers. Conversely, the thickness of the active layer should match the penetration length of incident light, which is typically between 80 and 200 nm for organic semiconductors.

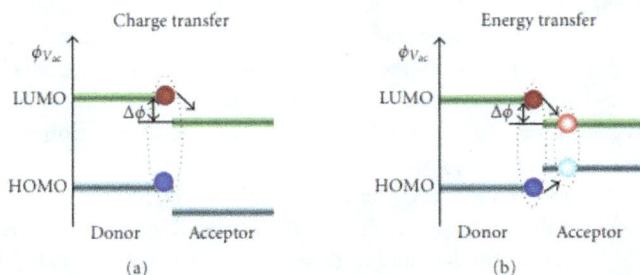

Figure 5: The interface between two different semiconducting polymers (D = donor, A = acceptor) can facilitate either charge transfer by splitting the exciton or energy transfer, where the whole exciton is transferred from the donor to the acceptor (Reproduced from Recent Advances in Organic Solar Cell, Hindawi publication, online journal) [28].

6. Morphology characterization and optimization

Exciton production, charge creation, and diffusion all take place in the photoactive layer of OSCs with bulk heterojunction [28]. As a result, techniques for monitoring and managing the shape of the active layer have been continually investigated. The observation of the active layer is dependent on the contrast of the blend film, whereas the regulation of the morphology is dependent on molecular self and hetero interactions, including the solvent, which are related to the molecular structures. Since non-fullerene acceptors (NFA) are developing so quickly, it is crucial to comprehend how morphological optimization and the associated molecular structures relate to one another. In this section, we will focus on the composition of the donor and acceptor phases and the effects of structures on molecular interaction and packing of the donor or acceptor. We mainly address the morphological change approach of the non-fullerene era. In the meanwhile, a brief overview of this section will be provided, focusing on the quantitative morphological standards that various characterization techniques satisfy.

6.1   Morphology characterization

Imaging and scattering are two universal characterization methods for bulk heterojunction. OM, AFM, PiFM, and TEM are the imaging tools that are easily accessible. These real-space data types are simple to understand intuitively, making them suitable as tools for basic morphological characterization. AFM is a better tool for observing surface roughness, PiFM is useful for differentiating between donors and acceptors in the BHJ active layer, and OM or TEM is frequently used to analyze composition or size distribution. By combining the contrast offered by electron energy loss spectroscopy with the energy-filtered TEM, model computations can be used to determine the quantitative composition of a sample. This technique is less common and more complicated. To evaluate how the donors and acceptors aggregate in their mix solutions, cryogenic transmission electron microscopy, or cryo TEM, has recently been used in addition to other characterization techniques for the active layer morphology of OSCs. In order to describe the morphology of the active layers, Raman spectroscopy was also utilized to assess the neat and blend films, especially for the ternary active layers. The compatibility of the materials used can be seen in the contact angle or Raman mapping photos. By utilizing RSoXS and GIWAXS to average thousands of domains, it is possible to determine the molecular packing and phase size distribution of polymer blends, which can provide statistically significant information. Again, elements of the device that cross over in the 2D projection have no impact on scattering. GIWAXS is an adequate method to describe molecular stacking and crystallinity in NFA-based film. As compared to the film employing fullerenes as acceptors, there is a significantly reduced contrast between the blended elements, which makes identifying phase composition particularly difficult. The decrease in contrast can be explained by the greater similarity between the structural motif ratio and C/H ratio of the donor and NFA. For this, the nanoscale awareness of local order and composition is provided by solid-state nuclear magnetic resonance (ss-NMR). Additionally beneficial and complementary would be the use of interfacial structures.

6.2   Morphology optimization

The optimization techniques will be divided into two groups in this section: thermodynamic optimization and kinetic optimization. We will also address current developments in the morphological control of the photoactive layers. Controlling interactions between self- and hetero-molecules (such as acceptor and donor) is one of the uses of thermodynamic techniques. Several controls for solution aggregation, film formation, and postfilm processing are included in kinetic techniques. The final film morphology can be changed by adjusting crystallization and phase separation through these parameters.

6.2.1 Thermodynamic optimization

We will split the thermodynamic optimization procedure into two stages in order to control interactions between homologous and heterologous molecules. The tiny molecules of the efficient NFAs have a strong tendency to combine into a stable thermodynamic state because of their distinctive fused ring conjugated structure. Therefore, modulating the intermolecular connections is an effective way to manage the molecular packing of acceptors. The use of non- covalent interactions into chemical structure design is a simple and useful strategy in this field. For example, studies from the literature indicate that higher molecular crystallinity than the counterpart (ITIC) may result from hydrogen bonding between acceptors with hydroxyl groups (Figure 6(a-c)).

Furthermore, compared to the molecule (IT-DOH) with hydroxyl groups on only one terminal, the molecule (IT-DOH) with hydroxyl groups on both terminals shows better molecular packing along the backbone and - directions, resulting in a maximum efficiency of 12.5% for PBDB-T-based devices that have not been annealed. Similarly, Zhan et al. were able to convert the two-dimensional molecular packing of the fused ring acceptor-INIC into a three-dimensional structure by introducing fluorine atoms into the end-group and benzene ring, respectively, as side chains.

The fluorine atom introduced interaction gives rise to four stable configurations in the fluorinated FINIC dimers, as the authors showed via DFT simulations. This is the foundation for the creation of three-dimensional stacking in its single crystals. While the fluorinated blends demonstrated improved face-on orientation and crystallinity, GIWAXS revealed that the matching polymer: NFA blends did not have a comparable three-dimensional structure. The improved dimer packing that came about as a result of fluorinating the side chains and end groups explain this. The spatial steric hindrance of side chains can affect the arrangement of molecules in addition to non-covalent interactions. It was found that the molecular orientation with respect to the substrate changes from edge-on to edge-off when the side chain changes from n-cetyl (85%) to 2hexyldecyl (88%), as illustrated in Figure 6d. From edge-on to face on, the molecule orientation with respect to the substrate changes. Similarly, previous studies show that PSFTZ crystallinity can be successfully controlled by platinum (II) complexation.

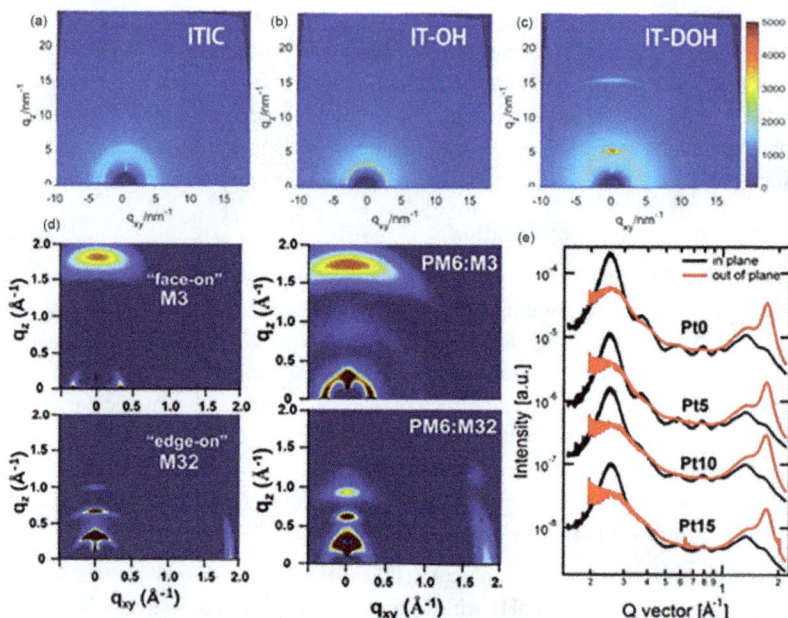

Figure 6. (a–c) GIWAXS images of ITIC, IT-OH, and IT-DOH films. (d) GIWAXS images of thin film based on neat M3, neat M32 and their blends with PM6. (e) The in-plane and out-of-plane line profiles of neat polymer films extracted from GIWAXS with different Pt ratio (Copyright @2022 Science China Chemistry; used with permission) [3].

Figure 6e showed how the platinum (II) complex's big benzene ring limits polymer aggregation strength and increases steric hindrance along the polymer main chain, both of which enhance device performance. The behavior of homologous molecules in terms of alignment, such as ordering, orientation, etc., is typically influenced by interactions. The miscibility of heterologous molecules will change when the interactions between them are altered. One of the forces causing phase separation is the donor and acceptor's miscibility, which usually has an effect on the domain's size, composition, etc. Recent work by Lee's group [39] has shown that the miscibility of the donor and acceptor also disturbs the uniformity of the morphology along the vertical direction. They found that the phase properties in thick films for the system with good miscibility are similar to those in thin films, while the systems with low miscibility show significant changes (Figure 7a).Therefore, using miscibility modulation to achieve a uniform morphological distribution in the vertical direction in thick films is promising. Thick film devices can be made efficiently with this technique. Ade and Ye et al.'s well-known works [40,41] demonstrate how miscibility controls phase separation and composition. They discovered a quantifiable relationship between the amorphous interaction parameter (χaa) and the filling factor (FF) of the OSCs.

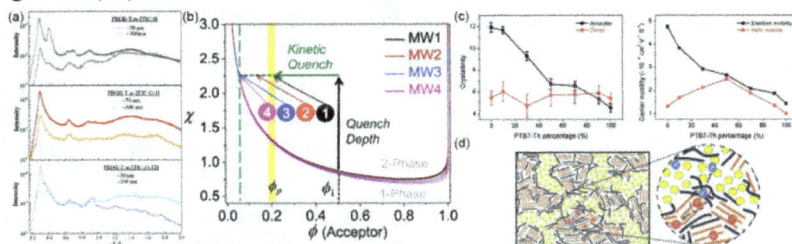

Figure 7. (a) GIWAXS scattering profiles along the in-plane direction for blends with different thickness. (b) Composition of mixed domain estimated from R-SoXS for PM6:IT-4F blends that needed to be quenched close to the percolation threshold for best performance. (c) The FOIC crystallinity quantified from GIWAXS and carrier mobility at different PTB7-Th content. (d) Illustration of the hierarchical morphology in BTR: NITI: PC71BM blends. The green circles represent PC71BM, the navy rods represent NITI and the orange rods represent BTR (Reproduced from Recent progress in organic solar cells (Part II device engineering), Science China chemistry, online journal) [3].

This model's relationship between the domain composition, FF, and (χaa) could provide the basis for simulation methods and miscibility testing, reducing the need for trial-and-error testing. Furthermore, they offered a framework that could be applied to accelerate the optimization of morphology in different systems by employing kinetic parameters such as quench depth and percolation threshold to guide morphological revolution. Interactions between heterologous molecules can become quite complex in ternary and quaternary systems. Rather of specifying a quantitative framework directly, assumptions are often made based on findings from physics, molecular packing, phase separation, or Hansen solubility parameter predictions. The following summary of empirical morphological optimization is still relevant. Furthermore, the study reveals that the highly crystalline polymer P1 unites the conventional PBDB-T:IT-M and PBDB-T:ITIC blends. The excellent compatibility between P1 and PBDB-T significantly enhanced the blended film's crystallinity, even with a low P1 level of 5%. This resulted in an optimal FF of the device that exceeded 78%. This technique has been widely applied in different systems to improve the morphology and performance of multi-component systems by optimizing the host morphology while taking compatibility and crystallinity into account. Lu et al. [43] discovered that the ITIC:ITIC-Th blends possess a concealed molecular packing structure, which is comparable to this. They demonstrated how the packing regularity of this structure is stronger than that of the two pure references, which improves device performance and carrier mobility. It's also helpful to employ a strong addition to display connections between the hosts and the guests. By altering this contact, the acceptor molecules' nucleation potential can be maximized, improving their crystallinity. In ternary blends, the balanced donor and acceptor crystallinity was achieved by managing the miscibility of the donors and acceptors. The authors found that in the PBDB-T:FOIC blade-coated film, PBDB-T and FOIC produced higher domain purity and had less compatibility. As a result, under blade-coating induction, the acceptor's crystallinity can be significantly greater than the donor's, causing the electron mobility to be larger than the total mobility. However, in the PTB7-Th:FOIC system, where the donor-acceptor compatibility is better than that of PBDB-T:FOIC blends, the improved crystallinity of the acceptor under scrape induction is not as evident. Based on the compatibility of these two systems and the characteristics of blade-induced crystallization, the scientists were able to establish balanced donor-acceptor crystallinity and carrier mobility by altering the ratio of PBDB-T:PTB7-Th. In addition to crystallinity optimization, a remarkable research case for phase separation optimization was presented.

They used the weakly crystalline small molecule acceptor NITI, the highly crystalline small molecule donor BTR, and the fullerene acceptor PC71BM. The novel aspect of this endeavor is how a hierarchical phase separation was effectively created by utilizing the differences in material compatibility. In particular, the high compatibility of BTR and NITI leads to a tendency for continuous mixing. However, the mismatch in crystallinity prevents the electrodes from effectively transporting and collecting the electrons in the NITI-dominated zones, which leads to the low FF and device performance of the system. Next, PC71 BM was presented by the authors. This compound has weak intermolecular connections and good compatibility with NITI, but significant intermolecular contacts and low compatibility with BTR. As a result, the ternary system experiences a hierarchical phase separation, with NITI:PC71BM acting as the electron transport channel and BTR:NITI acting as the D/A interface (Figure 7d). This is due to the fact that PC71BM tends to combine with NITI rather than BTR when intermolecular interactions are present. significant voltage, FF, and energy conversion efficiency are all simultaneously achieved by this profile by mitigating the negative effects of electron transport in NITI and avoiding the significant voltage loss caused by the BTR: PC71BM interface.

6.2.2 Kinetic optimization

In the part before this one, we looked at efficient ways to optimize morphology in binary and ternary systems from the perspective of intermolecular interaction. This section will discuss the morphological optimization from a kinetic standpoint during the film forming phase. Based on their qualities, solution-processed films will be divided into three categories: solution state, film formation state, and postfilm treatment. We will discuss the effects of each category on the crystallization and phase separation phases as well as the final morphology. The topic of controlling molecular pre-aggregation states in solutions was discussed. It should be emphasized that there are still difficulties in identifying molecular aggregation states at large concentrations, and that the current focus is on diluted solutions. Therefore, we limit our observations to empirical approaches for manipulating the pre-aggregation state in solutions. A crucial factor impacting the pre-aggregation phase is the solvent, which engages in diverse interactions with solute molecules. Using a variety of solutions, one typical method for modifying the structure of molecules—especially polymers—is to change their pre-aggregation condition. One of the most noteworthy achievements of this field of study is the regulation of a class of polymers exhibiting characteristics of temperature-dependent aggregation. By combining solution temperature control with a binary solvent system consisting of CB and DCB, they were able to regulate the aggregation state of PffBT4T-2OD. Furthermore, the aggregation state and film orientation of PffBT4T-C8C12 were modified through the application of 1,2,4-trimethylbenzene (TMB), a non-halogenated solvent, as shown in Figure 8a.

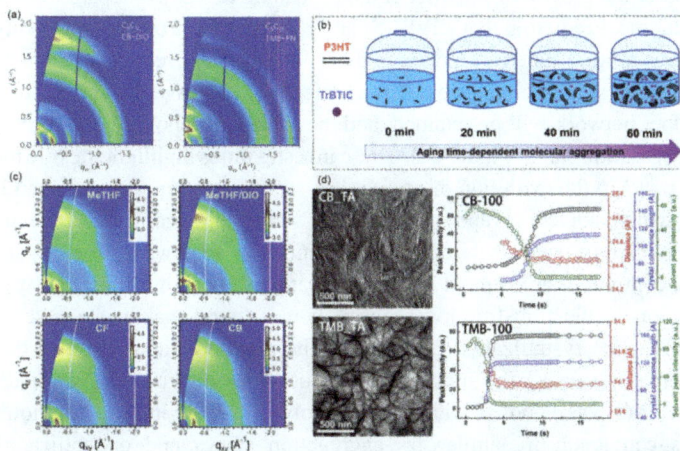

Figure 8. (a) GIWAXS images of PffBT4T-C8C12:PC71BM films processed with different solvents. (b) Illustration of the solution-ageing process. (c) GIWAXS images of PTzBI:N2200 blends under different solvent condition. (d) TEM images of PTzBI-Si:N2200 blends processed with CB and TMB and corresponding in-situ result ((Reproduced from Recent progress in organic solar cells (Part II device engineering), Science China chemistry, online journal) [3].

It has been widely used in nonfullerene systems to combine solvent and temperature management, which gives additional room for optimization in controlling the molecular pre-aggregation stages. For instance, by taking advantage of molecules' propensity to spontaneously combine in subpar solvents, one might enhance the crystallinity of molecules. A technique has been reported that alters the crystallization way and intensity in the film and largely affects the behavior of small molecules in solutions before they aggregate by varying the solvent's resting time. This work demonstrates the importance of the solution pre-aggregation state in the morphological development of thin films. To support the future structural evolution during film generation, in situ data must be used. However, the study also showed that under different fabrication temperatures, the change in evaporation rates caused by different boiling points significantly affects the blend shape, in addition to the solubility variation caused by the solvent. MeTHF effectively decreased the crystallization of N2200, as demonstrated in Figure 8c, when they assessed the morphology and device performance of PTzBI:N2200 blends treated with MeTHF, MeTHF/DIO, CF, and CB solvents. The pre-aggregation condition of N2200 in solution can actually be slightly improved by MeTHF, despite being a poor solvent for N2200. But since MeTHF has a low boiling point (80 °C), doing so will accelerate the formation of films and hinder the growth of polymer crystals, leading to a phase separation scale that is weakly crystalline. Predicting that PTzBI-Si and N2200 have strong compatibility and that a homogeneous mixing state can remain in the solution state, Liu and colleagues' research [44] also reached a similar conclusion. Thus, under MeTHF-processed conditions, a minor crystalline fiber network will be retained, and the mixed morphology can be frozen at the early phase separation stage (Figure 8d). These investigations highlight the role that solvent evaporation plays in the evolution of morphology. Because temperature has an impact on the rate at which solvent evaporates and the solute aggregation that occurs during processing, it can be said that temperature functions somewhat like a solvent. The advantage of continuous adjustment in the temperature field allows for greater opportunity for optimization in the film-forming process. It was found that in the case of slot-die coating, the kinetic crystallization process and the thermodynamic state of the solution could be altered by varying the substrate and solution temperatures. They said that even when the material was handled with different solvents, changing the solution and base temperatures can result in similar pre-aggregation stages and crystallization kinetic processes (Figure 9a, b). Comparable active layer geometry and device performance would result from this. The conclusion of this study is important for the scalable fabrication and optimization of OSCs. The geometry of the film-forming phase is, in fact, dependent on both phase separation behavior and solvent evaporation rate. Phase separation can be classified into three main categories: liquid-liquid (L-L), liquid-solid (L-S), and solid-solid (S-S). In general, the components that influence the phase separation mode are the

Third Generation Photovoltaic Technology   Materials Research Forum LLC
Materials Research Foundations 163 (2024) 52-91   https://doi.org/10.21741/9781644903032-3

composition of the solvent, donor, and acceptor. Janssen et al. [45] used in-situ optical measurements to show that polymer aggregation could start before L-L phase separation in PDPP5T-based fullerene systems due to binary solvent. Stated differently, the phase separation behavior in binary solvents is caused by the differing evaporation rates and solvent solubility qualities. This means that the architecture of the active layer can be improved by varying the kind and quantity of binary solvents (Figure 9 c). In order to optimize the P3HT:O-IDTBR system's active layer's crystallization and vertical phase separation, Han et al. [46] included a binary solvent CB:TCB. The authors showed that because of TCB's higher boiling point and better solubility for O-IDTBR than CB, the binary solvent was able to separate the crystallization processes of P3HT and O-IDTBR and lengthen the time needed for film manufacturing. The several crystallization processes that prevent cross-interference culminate in a highly crystalline layer.

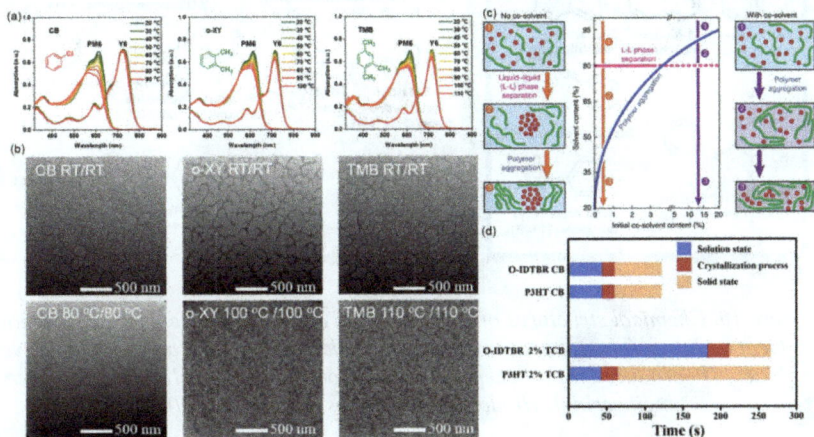

Figure 9: (a) Temperature-dependent UV-vis absorption spectra of PM6 and Y6 in CB, o-XY, and TMB. (b) TEM images of PM6:Y6 blends prepared by CB, o-XY, and TMB with different solution temperature. (c) The role of binary solvent in modulating polymer aggregation at higher solvent contents which, in turn, is expected to result in large-scale liquid-liquid phase separation during film drying. (d) The crystallization process of P3HT and o-IDTBR within film-forming period without and with TCB in CB solvent that is deduced from in-situ UV absorption spectra (Reproduced from Recent progress in organic solar cells (Part II device engineering), Science China chemistry, online journal) [28].

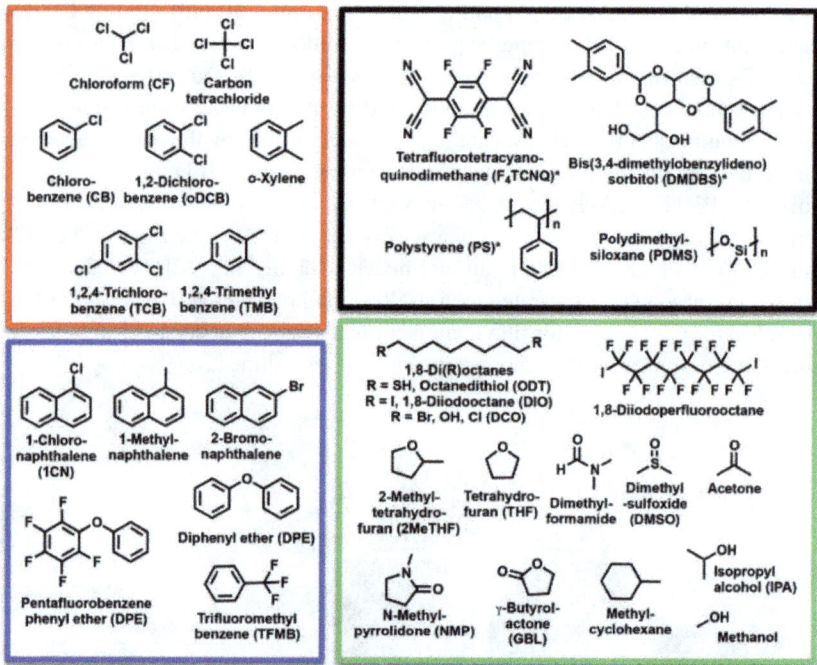

Figure 10. Chemical structures of normally used deposition solvents (red), aromatic solvent additives (blue), nonaromatic solvent additives (green), and solid additives (black) (Reproduced from Recent progress in organic solar cells (Part II device engineering), Hindawi publications, online journal) [3].

Once most of the solvent molecules have swept away from the film, the molecular chain segments have nearly totally lost their mobility. In this case, increasing molecular motion energy or decreasing molecular motion potential is the main objective of morphological optimization. Heat annealing or solvent evaporation annealing are often used procedures to optimize the shape and improve phase separation or crystallization of the blended film. The literature has a comprehensive explanation. Additives, such as solvent and solid additives, can effectively influence the form of the active layer during the film disintegration. The first solvent additive revealed that the P3HT:PC61BM blend film showed increased photocurrent upon the addition of n-octylthiol. Next, in order to further the phase separation of PCPDTBT: PC71BM blend films, n-octanedithiol was added, and the PCE was increased from 2.8% to 5.5%. According to Rogers et al.'s X-ray diffraction studies, the solvent addition may improve the donor polymers' crystalline structure. Further research may reveal the use of other 1,8-di(R)octane compounds (R can be thio-, chloro-, bromo-, iodo-, cyano-, or acetate) as solvent additives. As seen in Figure 10, todays widely used solvent additives include 1,8-diiodooctane (DIO), 1-chloronaphthalene (1CN), diphenyl ether (DPE), and others. Solid additives have also been successfully created in a similar manner. Due to the different properties of the host solvent and additives (such as boiling point and solubility capability), the solvent additives can generally significantly affect the phase morphology, including domain size, molecular crystallinity, and molecular orientation, which results in improved photovoltaic performance. Additives can further improve the photovoltaic performance of NFAs-based OSCs. Furthermore, two often utilized post-film treatment methods that might alter phase morphology are heat and solvent annealing. The photovoltaic performance of OSCs can also be improved by employing ternary or quaternary methods.

7. Semitransparent organic solar cells

Semitransparent organic solar cells (STOSCs) can be utilized as power generation windows for both electricity generation and light transmission. They consist of a semitransparent active layer with a structure of both top and bottom transparent electrodes. Due to their benefits of improving power conversion efficiency (PCE), strong processing performance, low cost, and mechanical flexibility, among others, organic solar cells (OSCs) have received a great deal of attention. Semitransparent organic solar cells (STOSCs) are typically composed of a semitransparent active layer with a structure that includes both top and bottom transparent electrodes. STOSCs have a powerful performance in building windows, car windows and greenhouse rooftops to meet the needs of people or plants.

STOSCs should offer a sizable PCE while preserving outstanding transparency to meet the objectives of this application. STOSCs should have a sizable PCE while yet performing exceptionally well in terms of transparency and color rendering for this application. Therefore, the top electrodes must have strong conductivity and exceptional transparency in the visible region. It was once a straightforward technique to use an ultrathin metal electrode as the top transparent electrode, however this single metal film has a tendency to develop an ad hoc island film, which results in subpar conductivity. A wide range of transparent electrode production techniques have been investigated to discover solutions to this issue, including the use of carbon nanotubes, graphene, transparent conductive oxide, metal nanowires, and transparent conductive polymers. Ag nanowires (Ag-NWs), the most representative metal nanowires, possess advantageous qualities like outstanding photo electronic performance and simple processing, and are frequently thought of as candidates for transparent electrodes. In contrast, the transparent conductive polymer poly(3,4-ethylenedioxythiophene):poly(styrenesulfonate) (PEDOT:PSS) has exceptional benefits such solution-process ability, low cost, and excellent transparency in the visible region. Regarding carbon nanotubes (CNTs), their high conductivity and transparency enable the development of STOSCs. Last but not least, roll-to-roll preparation of graphene, which has exceptional optical properties (transmittance close to 100%), is appropriate for the creation of large-scale OSCs.

The PCE and the region of light absorption are also significantly influenced by the active layer, which is likewise not insignificant. According to preliminary studies, the absorption spectra of fullerene-based OSCs are typically restricted to the range of 300 to 800 nm. With advantages like tunable energy levels, near-infrared light absorption, and long-term stability, non-fullerene small molecule acceptors redefine the range of the absorption spectrum of 300–1000 nm. This shows a significant improvement in STOSCs [47]. This has led to significant improvements in the PCE and average visible transmittance (AVT) of STOSCs based on non-fullerene acceptors. One of these technologies' greatest prospects is to be incorporated into energy-efficient buildings like windows and skylights. Although crystalline silicon-based modules now dominate this type of building integrated photovoltaic (BIPV), the opaque characteristic of silicon presents a rare opportunity for the adoption of developing photovoltaic contenders that can be rendered fully semi-transparent. These include systems based on amorphous silicon, kesterite, chalcopyrite, CdTe, dye-sensitized materials, organic materials, and perovskite. Due to its known, low-temperature production procedures, amorphous silicon has generally been the workhorse in the semi-transparent solar cell area. Due to the significant efficiency improvements demonstrated by these technologies, excitement has recently grown around alternate classes, particularly perovskites and the inorganic possibilities. Importantly, in the context of BIPV, each of them brings particular potential and difficulties. In addition to outlining the present state of research for all of the primary types of semi-transparent solar cell technologies, this topic review gives an overview of the larger advantages of using semi-transparent solar cells as built-in features.

8. Polymer-Based Solar Cells

Conjugated polymers make excellent materials for photovoltaic (PV) and inexpensive electronics. Recent studies show that polymer-based solar cells have power conversion efficiencies that are greater than 5%.2, 3, 4, 5, 6. Organics can be deposited by screen printing, doctor blading, inkjet printing, and spray deposition since these compounds may be made soluble. In order for polymer-based PV to compete on price with current grid power, high-throughput roll-to-roll processing is required. Additionally, flexible electronic devices can be built on plastic substrates because all of these deposition processes take place at low temperatures. Flexibility and light weight, in addition to the inherent economics of high-throughput manufacturing, are claimed to be able to reduce PV panel costs through reducing installation costs. Flexible PV also facilitates specific markets for things like transportable power generation and striking structural design possible.

The construction of various organic electronics parts, such as field-effect transistors, light-emitting diodes, memory cells, solar cells, and sensors, using organic materials with semiconductor properties has recently been the subject of considerable research. Inorganic semiconductors are being replaced by conjugated polymers as a result of the invention of conductive polyacetylene. Meaningful advancement in the manufacture of organic solar cells (OSCs) has become feasible since the discovery of the most recent generation of conductive conjugated polymers, which are utilized to make field-effect transistors and current light-emitting diodes. These polymers are versatile, come in a variety of forms, and have excellent mechanical qualities. distinct layers of donor and acceptor photoactivators in a binary structure Because of their high optical absorption coefficient, they can be employed as very thin films with a thickness of about 100 nm. In addition to being able to deposit films from solutions at normal pressure onto flexible substrates of a wide area, the advantages mentioned above also make it possible to build an OSC using relatively affordable methods like inkjet printing and stamping technologies. Despite these benefits of employing polymers, the need for protective encapsulation from external effects and the relatively low power conversion efficiency (PCE) of 6-7% prevent the commercialization of the OSC. Almost all known types of organic photovoltaic cells can be divided into two main groups. Batteries with a binary structure and distinct layers of donor and acceptor photoactive components fall under the first group.

Batteries that only have one photoactive layer composed of a donor and an acceptor or that have a bulk heterojunction fall into the second category. The two conducting electrode layers, one of which is transparent to incident light, must be positioned between the active layers of polymer solar cells. For this purpose, typically an indium tin oxide (ITO) coating—a mixture of indium and tin oxides—is applied on a glass or flexible polymer substrate. The ITO layer is also covered with a layer of poly (3,4-ethylenedioxythiophene): poly(styrene sulfonate, or PEDOT: PSS), a conductive polymer used to transmit holes. As a result of better energy level matching between the electrode yield and the ITO's highest occupied molecular orbital (HOMO) level, this film also smooths the ITO's surface, removes surface shunts, and boosts the effectiveness of the entire collecting process. On the side of the active layer opposite it, a low work function metal electrode is utilized.

In principle, this is an aluminum electrode that can be further modified to increase the efficiency of solar cells by including a thin layer (less than 1 nm) of LiF underneath it [50]. Such an element is lighted by sunlight coming in from the side of a clear glass or polymer

Third Generation Photovoltaic Technology   Materials Research Forum LLC
Materials Research Foundations 163 (2024) 52-91   https://doi.org/10.21741/9781644903032-3

substrate. When radiation is absorbed by the active polymer or composite layer, electron-hole pairs (exciton) are formed. These exciton later disintegrate into electrons and holes gathered on opposing electrodes.

8.1   Fullerene-containing polymers for organic solar cell

Bulk fullerene polymer in academic labs, heterojunction solar cells are being studied as a type of solar cell. In an area where silicon-based inorganic solar cells predominate, polymeric organic-based photovoltaic cells have shown potential. A polymer-fullerene based photovoltaic cell is created when fullerene derivatives specifically function as electron acceptors for donor materials like P3HT (poly-3-hexyl thiophene-2,5-diyl).

The flexibility of organic solar cells, which enables their application to a larger range of surfaces, is just one of their many advantages over inorganic solar cells. Additionally, they may be produced much more rapidly and simply via spray deposition or inkjet printing, which results in a significant reduction in production costs. However, because they are not crystalline (like silicon), but rather are produced in a purposely disordered mixture of electron-acceptor and -donor materials (hence the name bulk heterojunction), they have a poorer efficiency of charge transfer.

However, efficiency of these new solar cell kinds has increased to more than 10%, from 2.5% in 2001, 5% in 2006, and above 10% in 2011. This is because improved technology for component solution processing enables more efficient blending of the acceptor and donor materials. Future solar cells made of polymer fullerene may have efficiency levels that are comparable to those of today's inorganic photovoltaic cells.

When donor molecules are photo excited, the electrons move from the HOMO to the LUMO energy state. Now that they are in the LUMO energy level, the electrons can go to neighboring acceptor molecules that are more electronegative and have lower energies. The transfer of electrons between the donor and acceptor is driven by the contrast in LUMO energy levels between them.

Fullerene molecules have recently been the focus of extensive and competitive study, and their incorporation into polymer chains as photo- and electro active moieties could result in the development of novel materials with distinctive structural, electrochemical, and photo physical properties. Recent years have seen the publication of numerous papers that make extensive use of the metathesis technique to acquire materials for solar cells. For instance, Eo et al. [48, 49] proposed the synthesis of vinyl-type polynorbornenes, a common electron-withdrawing component of the active layer in organic photovoltaic cells, whose structure incorporates fragments of (60)PCBM.

These polymers were used to create photovoltaic cells, where the fullerene-containing copolymer served as the active layer's n-type semiconductor. A number of studies that used a Grubbs catalyst to metathesis polymerize fullerene-containing monomers (FCMs) and evaluated the results in solar cells are also interesting. In the section of their research, a new novel norbornene-type monomer into fullerene-containing polymers and copolymers was synthesized in the presence of the first-generation Grubbs catalyst [(PCy3)2Cl2RuCHPh] [Figure.11].

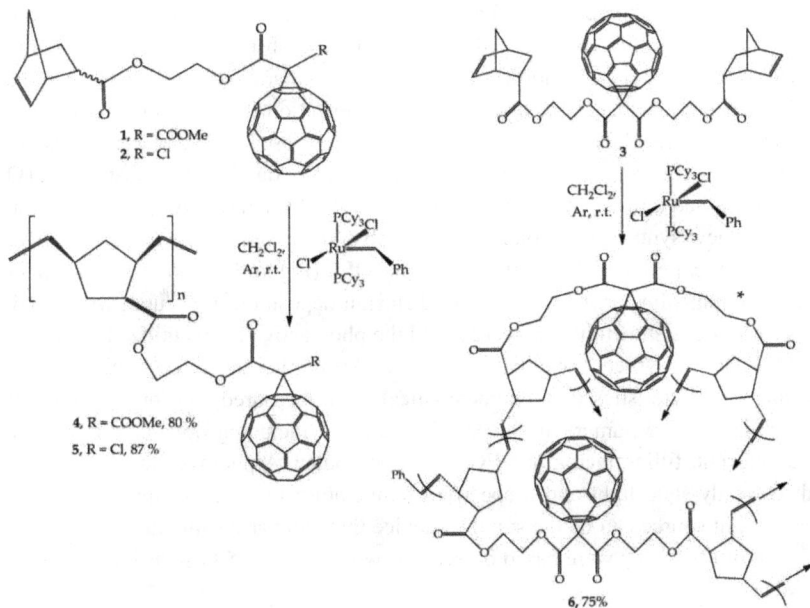

Figure.11 Ring-opening metathesis polymerization of fullerene-containing norbornene monomers (Reproduced from reference [51]).

9. Organic solar cells based on thin polymer films

A Knudsen effusion cell was utilized to create thin films of Polyanilines (PANIs) and polymers incorporating fullerenes using a vacuum deposition process [52]. The cylindrical cell's internal diameter was 4 mm, its length was 25 mm, and its operating temperature ranged from 500 to 650 K. Fullerene-containing monomers (FCMs) were thermally heated during deposition, which caused them to polymerize. Using the spin coating method, certain thin films were created from a solution of monomers containing fullerene. All of the films that were produced had uniform thickness and a conductivity of between 0.1 and 1.0 mS/cm. The temperature parameters of deposition from the Knudsen cell were chosen to increase the conductivity of polyaniline layers. The best results were obtained at temperatures between 500 and 550 K. Additionally, the freshly made films were protonated in hydrochloric acid solution vapors. As a result, a conductivity value of 1.0 mS/cm was attained for PANI films.

AFM pictures captured by a Nano Scan 3D were analyzed in order to monitor the surface quality and thickness of the deposited films. The photoactive layers' thickness fluctuated and took on values between 100 and 200 nm. It should be observed that excessive film thickness causes exciton recombination and lowers charge separation effectiveness. On the other hand, in overly thin sheets, the incident photon absorption and the amount of exciton generated decrease. The donor-acceptor polymer systems used in the test samples for organic solar cells were produced on a glass substrate with conductive and transparent ITO layers. ITO layers had a resistance of about of 10 $\Omega/\square$. In this study, PANI, conventional fullerene, and a new synthesized monomer monosubstituted methanofullerene derivative were employed for experimental constructions of the OSC (Figure 12a and b). As the upper electrode, aluminum films created by thermo-diffusion deposition in vacuum were used. The OSC's structure is shown in Figure 12c, and the photoactive layers utilized were thin films of PANI and polymers containing fullerene. All of the generated OSC samples' current-voltage characteristics (CV characteristics) were measured, and on the basis of those measurements, the numerical values of parameters including open-circuit voltage short-circuit current, filling factor, and PCE were computed. A photovoltaic cell is often subjected to steady-state light and a specified temperature in order to measure its CV properties. A light source can be the sun or a device that simulates sunlight. Estimates of the coefficient of efficiency were based on settings with an AM 1.5 G solar intensity of P 0 = 1000 W/m2.

These parameters appeared to have values of Jsc = 0.6-1.8 mA/cm2 (short-circuit current), Voc = 0.6-0.8 V (open-circuit voltage), and FF = 0.6-0.8 (filling factor) for the various OSC experimental structures investigated in this work. For the examined organic solar cells, the greatest PCE values were around 2%. These numbers were derived for the derivatives of methanofullerene-based structures.

Thus, it was shown that the production of OSC based on binary donor-acceptor systems requires the mixing of PANI with polymers containing fullerene. The solar cells under investigation here are unique from those studied previously [53] in that they can be manufactured on flexible substrates.

Materials Research Forum LLC
https://doi.org/10.21741/9781644903032-3

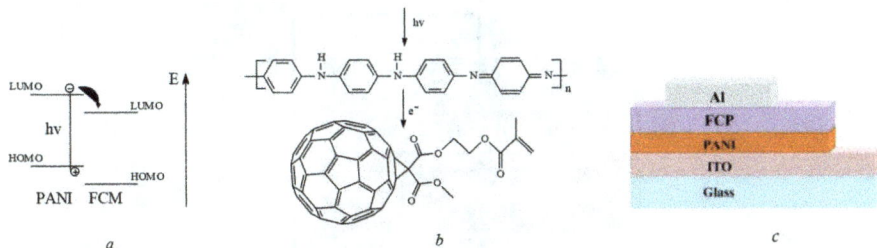

Figure 12. (a) An energy level diagram of the PANI/FCM system; (b) process of photon absorption and charge separation in this structure; (c) multilayer film structure of OSc (Reproduced from reference [51]).

9.1 Polymerizable methanofullerene as a buffer layer material for organic solar cell

New fullerene derivative-based (n-type) and electron-conjugated polymer-based (p-type) semiconductor material combinations are actively being developed globally in recent years. The use of charge-selective buffer layers is thought to be the only way to produce organic solar cells with a high efficiency of light conversion [54]. PEDOT: PSS and a number of inorganic oxides are often used materials for creating these layers. Because PEDOT: PSS has acidic characteristics, using it shortens the time that solar cells are operational. At the same time, materials in the photoactive layer exhibit oxidising qualities that make them more brittle. These metal oxides in high oxidation states (MoO_3, V_2O_5, and WO_3). Even with relatively inert titanium dioxide TiO_2, the issue is seen [55]. Inverted configuration organic solar cells with much higher operational stability and low active metal content offer the best chances for practical deployment. However, the development of selective electron-transport buffer layers (ETL) based on n-type semiconductor materials is necessary to produce these devices. We have created inverted solar cells with an ITO cathode, an ETL (electro-transport layer) that contains fullerene, a photoactive layer, a hole-transporting layer made of MoO_3, and an Ag anode (Figure13).

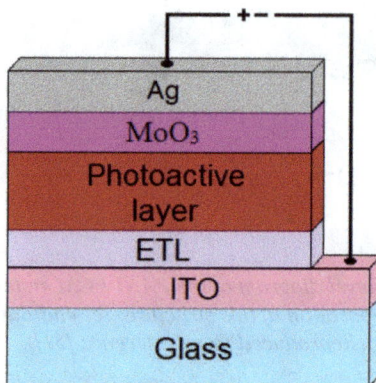

Figure 13. Schematic architecture of an inverted organic solar cell (Reproduced from reference [51]).

10. Challenges and outlook

Efficiency and price are the two key problems with the different solar cells. In the case of a monocrystalline cell, the crystallization process is quite expensive. Similar crystallization processes occur in polycrystalline silicon cells; however, these cells have structural problems. Developments in photovoltaic technology are crucial for addressing a number of solar panel-related issues, such as instability and deterioration, power conversion efficiency, durability, and manufacturing costs. Low installation and maintenance costs, noise and air pollution freedom, little reliance on non-renewable natural resources, good return on investment, and decreased greenhouse gas emissions are some of the main difficulties of photovoltaic. Potentially Induced Degradation Effect, Micro Cracks, Snail Trails, Electrical difficulties, Roof Issues, Birds, Inverter Faults, Hotspots, Delamination, and Internal Corrosion are a few common difficulties with solar panels.

OSCs have shown tremendous promise for enhancing device productivity and long-term operational stability. Both the device's performance and transmittance should be adjusted in order to produce high-performance ST-OSCs. Semi-transparent organic solar cells (STOSCs) have received a lot of attention recently because of their potential for use in building-integrated photovoltaic (BIPV) and the internet of things (IoT). Organic photovoltaic materials have tailorable and discontinuous absorption characteristics due to the structure of molecular orbitals, which gives them a distinct advantage for ST-OSCs. For the purpose of supplying electricity, partial photons can be trapped by ST-OSCs, whereas for daily needs, partial ones can pass through them. Power conversion efficiency

(PCE) and average visible transmittance are the two main metrics used to assess ST-OSC performance. The achieved light conversion efficiency suggests potential for this research to advance. Higher values of OSC efficiency will be achieved by properly selecting the organic material composition and optimizing the technological conditions for the manufacturing of thin films.

Conclusion

OSCs have demonstrated significant advantages in the search for low production costs while maintaining respectable power conversion efficiency. Some benefits offered by OSCs include the ability to modify the chemical characteristics, light weight, and simple construction. Although, many developments have been made towards the realization of OSCs, and inorganic molecules-based solar cells have demonstrated improved power conversion efficiency and environmental stability. By utilizing highly efficient absorber materials and inverted device structures in organic solar cells with tandem architecture, it is possible to achieve great performance in OSCs. In OSC, polymers are typically used as the absorber layer to improve device performance and light harvesting efficiency. The development of novel polymers with tiny band gaps and proper energy level arrangements is essential for achieving the highly effective OSC.

The use of this truly green energy, solar energy, has emerged as one of the most hotly debated issues in both the government and scientific communities as the global energy crisis and environmental problem become more serious. The significant financial and human resources committed to this sector will undoubtedly improve the performance of polymer organic solar cells. The polymer organic solar cell has the potential to be very successful in the future, with a power conversion efficiency of 4.4% already achieved and plenty of space for improvement. To make future advancements in polymer organic solar cells, attention should be given to the following factors:

1. Screening low band-gap donor materials is the first step in increasing sunlight harvesting since these materials' absorption range better spans the solar spectrum.

2. Enhancing the electrodes to produce better electric contacts at both electrode interfaces for improved charge carrier collecting.

3. Changing the shape of the composite films to create more orderly, continuous paths for charge carriers to travel rapidly and directly to the electrodes.

4. Introducing innovative device architectures. For example, devices with several junctions and layers are better suited to utilizing incident sunlight.

References

[1] Abodunrin, T. J., Boyo, A. O., Usikalu, M. R. & Kesinro, O. (2018). Spectral responses of B.vulgaris dye-sensitized solar cells to change in electrolyte, IOP Conf. Series: Earth and Environmental Science, 173: 012047. https://doi.org/10.1088/1755-1315/173/1/012047

[2] Tsokos, K. A. (28 January 2010). Physics for the IB Diploma Full Colour. Cambridge University Press. ISBN 978-0-521-13821-5.

[3] Liu Y, Liu B, Ma CQ, Huang F, Feng G, Chen H, Hou J, Yan L, Wei Q, Luo Q, Bao Q, Ma W, Liu W, Li W, Wan X, Hu X, Han Y, Li Y, Zhou Y, Zou Y, Chen Y, Li Y, Chen Y, Tang Z, Hu Z, Zhang ZG, B Z. Science China Chemistry, 2022, 65: 224-26. https://doi.org/10.1007/s11426-021-1180-6

[4] Chiang CK, Fincher CR, Park YW Jr, Heeger AJ, Shirakawa H, Louis EJ, Gau SC, MacDiarmid AG. Electrical conductivity in doped polyacetylene. Physical Review Letters. 1977;39:1098-1101. DOI: 1010.1103/PhysRevLett.39.1098. https://doi.org/10.1103/PhysRevLett.39.1098

[5] Shaheen SE, Ginley DS, Jabbour GE. Organic-based photovoltaics: Toward low-cost power generation. MRS Bulletin. 2005;30:10-19. https://www.calpoly.edu/~rechols/Phys422/ MRS2005Intro.pdf. https://doi.org/10.1557/mrs2005.2

[6] C. W. Tang and A. C. Albrecht, "Photovoltaic effects of metal−chlorophyll-a−metal sandwich cells," The Journal of Chemical Physics, vol. 62, no. 6, pp. 2139-2149, 1975. https://doi.org/10.1063/1.430780

[7] C. W. Tang, "Two-layer organic photovoltaic cell," Applied Physics Letters, vol. 48, no. 2, pp. 183-185, 1986. https://doi.org/10.1063/1.96937

[8] J. Xue, S. Uchida, B. P. Rand, and S. R. Forrest, "4.2% efficient organic photovoltaic cells with low series resistances," Applied Physics Letters, vol. 84, no. 16, pp. 3013-3015, 2004. https://doi.org/10.1063/1.1713036

[9] J. Xue, S. Uchida, B. P. Rand, and S. R. Forrest, "Asymmetric tandem organic photovoltaic cells with hybrid planar-mixed molecular heterojunctions," Applied Physics Letters, vol. 85, no. 23, pp. 5757-5759, 2004. https://doi.org/10.1063/1.1829776

[10] M. Reyes-Reyes, K. Kim, and D. L. Carroll, "High-efficiency photovoltaic devices based on annealed poly(3- hexylthiophene) and 1-(3-methoxycarbonyl)-propyl-1-

phenyl-(6,6)C61 blends," Applied Physics Letters, vol. 87, no. 8, Article ID 083506, 3 pages, 2005. https://doi.org/10.1063/1.2006986

[11] J. Xue, B. P. Rand, S. Uchida, and S. R. Forrest, "Mixed donoracceptor molecular heterojunctions for photovoltaic applications. II. Device performance," Journal of Applied Physics, vol. 98, no. 12, Article ID 124903, 9 pages, 2005. https://doi.org/10.1063/1.2142073

[12] C. W. Tang and A. C. Albrecht, "Photovoltaic effects of metal−chlorophyll-a−metal sandwich cells," The Journal of Chemical Physics, vol. 62, no. 6, pp. 2139-2149, 1975. https://doi.org/10.1063/1.430780

[13] C. W. Tang, "Two-layer organic photovoltaic cell," Applied Physics Letters, vol. 48, no. 2, pp. 183-185, 1986. https://doi.org/10.1063/1.96937

[14] J. Xue, S. Uchida, B. P. Rand, and S. R. Forrest, "4.2% efficient organic photovoltaic cells with low series resistances," Applied Physics Letters, vol. 84, no. 16, pp. 3013-3015, 2004. https://doi.org/10.1063/1.1713036

[15] J. Xue, S. Uchida, B. P. Rand, and S. R. Forrest, "Asymmetric tandem organic photovoltaic cells with hybrid planar-mixed molecular heterojunctions," Applied Physics Letters, vol. 85, no. 23, pp. 5757-5759, 2004. https://doi.org/10.1063/1.1829776

[16] K. Takahashi, N. Kuraya, T. Yamaguchi, T. Komura, and K. Murata, "Three-layer organic solar cell with high-power conversion efficiency of 3.5%," Solar Energy Materials and Solar Cells, vol. 61, no. 4, pp. 403-416, 2000. https://doi.org/10.1016/S0927-0248(99)00163-4

[17] K. Takahashi, N. Kuraya, T. Yamaguchi, T. Komura, and K. Murata, "Three-layer organic solar cell with high-power conversion efficiency of 3.5%," Solar Energy Materials and Solar Cells, vol. 61, no. 4, pp. 403-416, 2000. https://doi.org/10.1016/S0927-0248(99)00163-4

[18] A. J. Breeze, A. Salomon, D. S. Ginley, H. Tillmann, H. Horhold, and B. A. Gregg, "Improved efficiencies in polymer-perylene diimide bilayer photovoltaics," in Organic Photovoltaics III, vol. 4801 of Proceedings of SPIE, pp. 34-39, Seattle, Wash, USA, June 2002. https://doi.org/10.1117/12.452436

[19] A. J. Breeze, A. Salomon, D. S. Ginley, B. A. Gregg, H. Tillmann, and H.-H. Horhold, "Polymer-perylene diimide het- " erojunction solar cells," Applied Physics Letters, vol. 81, no. 16, pp. 3085-3087, 2002. https://doi.org/10.1063/1.1515362

[20] J.-I. Nakamura, C. Yokoe, K. Murata, and K. Takahashi, "Efficient organic solar cells by penetration of conjugated polymers into perylene pigments," Journal of Applied Physics, vol. 96, no. 11, pp. 6878-6883, 2004. https://doi.org/10.1063/1.1804245

[21] H. Sirringhaus, "Device physics of solution-processed organic field-effect transistors," Advanced Materials, vol. 17, no. 20, pp. 2411-2425, 2005 https://doi.org/10.1002/adma.200501152

[22] Chiang CK, Fincher CR, Park YW Jr, Heeger AJ, Shirakawa H, Louis EJ, Gau SC, MacDiarmid AG. Electrical conductivity in doped polyacetylene. Physical Review Letters. 1977;39:1098-1101. DOI: 1010.1103/PhysRevLett.39.1098. https://doi.org/10.1103/PhysRevLett.39.1098

[23] Shaheen SE, Ginley DS, Jabbour GE. Organic-based photovoltaics: Toward low-cost power generation. MRS Bulletin. 2005;30:10-19. https://www.calpoly.edu/~rechols/Phys422/ MRS2005Intro.pd. https://doi.org/10.1557/mrs2005.2

[24] Kesinro, R. O., Boyo, A. O., Akinyemi, M. L. & Mola, G. T. (2019). Fabrication of P3HT: PCBM bulk heterojunction organic solar cell, IOP Conf. Series: Earth and Environmental Science, 331: 01202. https://doi.org/10.1088/1755-1315/331/1/012028

[25] Kalyani, N. T. & Dhoble, S. J (2018). Empowering the future with organic solar cell devices, Nanomaterials for Green Energy.

[26] Kesinro, R. O., 2 Boyo, A. O, 1 Akinyemi, M. L., 1,3Emetere, M. E., 1 Aizebeokhai, A. P , Progress on Organic Solar Cells: A Short Review, 2021 IOP Conf. Ser.: Earth Environ. Sci. 665 012036. https://doi.org/10.1088/1755-1315/665/1/012036

[27] C. K. Chiang, C. R. Fincher Jr., Y. W. Park, et al., "Electrical conductivity in doped polyacetylene," Physical Review Letters, vol. 30, no. 17, pp. 1098-1101, 1977. https://doi.org/10.1103/PhysRevLett.39.1098

[28] Thomas Kietzk, Recent Advances in Organic Solar Cell Volume 2007, Article ID 40285, 15 pages, Advances in OptoElectronics,, Hindawi Publishing Corporation. doi:10.1155/2007/40285. https://doi.org/10.1155/2007/40285

[29] T. Kietzke, D. A. M. Egbe, H.-H. Horhold, and D. Ne, "Comparative study of M3EH-PPV-based bilayer photovoltaic devices," Macromolecules, vol. 39, no. 12, pp. 4018- 4022, 200. https://doi.org/10.1021/ma060199l

[30] M. M. Koetse, J. Sweelssen, K. T. Hoekerd, et al., "Efficient polymer: polymer bulk heterojunction solar cells," Applied Physics Letters, vol. 88, no. 8, Article ID 083504, 3 pages, 2006. https://doi.org/10.1063/1.2176863

[31] Chang CY, Tsai FY. Highly efficient red fluorescent dyes for organic light-emitting diodes, Journal of Material Chemistry, 2011, 21: 5710-5715 https://doi.org/10.1039/c0jm03109g

[32] Distler A, Sauermann T, Egelhaaf HJ, Rodman S, Waller D, Cheon KS, Lee M, Guldi DM., The Effect of PCBM Dimerization on the Performance of Bulk Heterojunction Solar Cells, Advanced Energy Material, 2014, 4: 1300693. https://doi.org/10.1002/aenm.201400171

[33] Yan L, Yi J, Chen Q, Dou J, Yang Y, Liu X, Chen L, Ma CQ. Ultrafast Growth of High-Quality Monolayer WSe2 on Au Material Chemistry A, 2017, 5: 10010-10020. https://doi.org/10.1039/C7TA02492D

[34] S. Holliday, R. Ashraf, A. Wadsworth, D. Baran, S. Yousaf, C. B. Nielsen, C.-H. Tan, S. D. Dimitrov, Z. Shang, N. Gasparini, M. Alamoudi, F. Laquai, C. J. Brabec, A. Salleo, J. R. Durrant and I. McCulloch, Nat. Commun., 2016, 7, 11585. https://doi.org/10.1038/ncomms11585

[35] J. Cai, H. Wang, X. Zhang, W. Lia, D. Lia, Y. Mao, B. Dua, M. Chen, Y. Zhuang, D. Liu, H.-L. Qin, Y. Zhao, J. A. Smithe, R. C. Kilbridee, A. J. Parnelle, R. A. L. Jonese, D. G. Lidzeye and T. Wang, Fluorinated solid additives enable high efficiency non-fullerene organic solar cells Journal of Materials Chemistry A, 2020, 8, 4230-4238. https://doi.org/10.1039/C9TA13974E

[36] S. Park and H. J. Son, Intrinsic photo-degradation and mechanism of polymer solar cells: the crucial role of non-fullerene acceptors, Journal of Materials Chemistry. A, 2019, 7, 25830- 258. https://doi.org/10.1039/C9TA07417A

[37] N. Y. Doumon, M. V. Dryzhov, F. V. Houard, V. M. Corre, A. Chatri, P. Christodoulis and L. Koster, Photo stability of Fullerene and Non-Fullerene Polymer Solar Cells: The Role of the Acceptors, Applied Material Interfaces, 2019, 11, 8310-8318. https://doi.org/10.1021/acsami.8b20493

[38] Dai S, Zhou J, Chandrabose S, Shi Y, Han G, Chen K, Xin J, Liu K, Chen Z, Xie Z, Ma W, Yi Y, Jiang L, Hodgkiss JM, Zhan X, High-Performance Fluorinated Fused-Ring Electron Acceptor with 3D Stacking and Exciton/Charge Transport, Advanced Materials, 2020, 32: 2000645. https://doi.org/10.1002/adma.202000645

[39] Lee S, Park KH, Lee J, Back H, Sung MJ, Lee J, Kim J, Kim H, KimY, Kwon S, Lee K. Achieving Thickness-Insensitive Morphology of the Photoactive Layer for Printable Organic Photovoltaic Cells via Side Chain Engineering in Nonfullerene Acceptors, Advanced Energy Materials, 2019, 9: 1900044. https://doi.org/10.1002/aenm.201900044

[40] Ye L, Hu H, Ghasemi M, Wang T, Collins BA, Kim JH, Jiang K, Carpenter JH, Li H, Li Z, McAfee T, Zhao J, Ch en X, Lai JLY, Ma T, Bredas JL, Yan H, Ade H, Quantitative relations between interaction parameter, miscibility and function in organic solar cells, Nature Materials, 2018, 17: 253-260 https://doi.org/10.1038/s41563-017-0005-1

[41] Ye L, Li S, Liu X, Zhang S, Ghasemi M, Xiong Y, Hou J, Ade H. Joule, Quenching to the Percolation Threshold in Organic Solar Cells, 2019, 3: 443-45. https://doi.org/10.1016/j.joule.2018.11.006

[42] Mai J, Xiao Y, Zhou G, Wang J, Zhu J, Zhao N, Zhan X, Lu X. Hidden Structure Ordering Along Backbone of Fused-Ring Electron Acceptors Enhanced by Ternary Bulk Heterojunction, Advanced Materials, 2018, 30: 1802888. https://doi.org/10.1002/adma.201802888

[43] R, Yao H, Hong L, Qin Y, Zhu J, Cui Y, Li S, Hou Design and application of volatilizable solid additives in non-fullerene organic solar cells. Nature Communications, 2018, 9: 4645. https://doi.org/10.1038/s41467-018-07017-z

[44] Liu Y, Liu B, Ma CQ, Huang F, Feng G, Chen H, Hou J, Yan L, Wei Q, Luo Q, Bao Q, Ma W, Liu W, Li W, Wan X, Hu X, Han Y, Li Y, Zhou Y, Zou Y, Chen Y, Liu Y, Meng L, Li Y, Chen Y, Tang Z, Hu Z, Zhang ZG, Bo Z., Recent progress in organic solar cells (Part II device engineering). Sci China Chem, 2022, 65, https://doi.org/10.1007/s11426-022-1256-8. https://doi.org/10.1007/s11426-022-1256-8

[45] Jacob van Franeker JJ, Turbiez M, Li W, Wienk MM, Janssen RAJ., A real-time study of the benefits of co-solvents in polymer solar cell processing. Nature Communications, 2015, 6: 622. https://doi.org/10.1038/ncomms7229

[46] Liang Q, Jiao X, Yan Y, Xie Z, Lu G, Liu J, Han Y., Separating Crystallization Process of P3HT and O-IDTBR to Construct Highly Crystalline Interpenetrating Network with Optimized Vertical Phase Separation, Advanced Functional Materials, 2019, 29: 180759. https://doi.org/10.1002/adfm.201807591

[47] Yan L, Yi J, Chen Q, Dou J, Yang Y, Liu X, Chen L, Ma CQ., J External load-dependent degradation of P3HT:PC61BM solar cells: behaviour, mechanism, and

Materials Research Forum LLC
https://doi.org/10.21741/9781644903032-3

method of suppression, Journal of Material Chemistry A, 2017, 5: 10010-10020. https://doi.org/10.1039/C7TA02492D

[48] Eo M, Han D, Park M, Hong M, Do Y, Yoo S, Lee M. Polynorbornenes with pendant PCBM as an acceptor for OPVs: Ring-opening metathesis versus vinyl-addition polymerization. European Polymer Journal. 2014;5:37-44. DOI: 10.1016/j.eurpolymj.2013.11.018. https://doi.org/10.1016/j.eurpolymj.2013.11.018

[49] Eo M, Lee S, Park M, Lee M, Yoo S, Do Y. Vinyl-type polynorbornenes with pendant PCBM: A novel acceptor for organic solar cells. Macromolecular Rapid Communications. 2012;33:1119-1125. DOI: 10.1002/marc.201200023. https://doi.org/10.1002/marc.201200023

[50] Markov DE, Hummelen JC, Blom PWM, Sieval AB. Dynamics of exciton diffusion in poly(p-phenylene vinylene) fullerene heterostructures. Physical Review B. 2005;72:045216. DOI: doi.org/10.1103/PhysRevB.72.045216. https://doi.org/10.1103/PhysRevB.72.045216

[51] Renat B. Salikhov, Yuliya N. Biglova and Akhat G. Mustafin, New Organic Polymers for Solar Cells. http://dx.doi.org/10.5772/intechopen.74164. https://doi.org/10.5772/intechopen.74164

[52] Salikhov RB, Biglova YN, Yumaguzin YM, Salikhov TR, Miftakhov MS, Mustafin AG. Solar-energy photoconverters based on thin films of organic materials. Technical Physics Letters. 2013;39:854-857. https://link.springer.com/article/10.1134/S1063785013100106. https://doi.org/10.1134/S1063785013100106

[53] Wang W, Schiff EA. Polyaniline on crystalline silicon heterojunction solar cells. Applied Physics Letters. 2007; 91:133504. DOI: 10.1063/1.2789785. https://doi.org/10.1063/1.2789785

[54] Yang P, Chen S, Liu Y, Xiao Z, Ding L. A pyridine-functionalized pyrazolinofullerene used as a buffer layer in polymer solar cells. Physical Chemistry Chemical Physics. 2013;15:17076-17078. DOI: 10.1039/C3CP53426J. https://doi.org/10.1039/c3cp53426j

[55] Kim D, Jeong M, Seo H, Kim Y. Oxidation behavior of P3HT layers on bare and TiO2 - covered ZnO ripple structures evaluated by photoelectron spectroscopy. Physical Chemistry Chemical Physics. 2015;17:599-604. DOI: 10.1039/C4CP03665D. https://doi.org/10.1039/C4CP03665D

Third Generation Photovoltaic Technology
Materials Research Foundations 163 (2024) 92-117

Materials Research Forum LLC
https://doi.org/10.21741/9781644903032-4

Chapter 4

Rational Design and Development of Copper Zinc Tin Sulfide Solar Cells

K.S. Rajni[1*], Narayanan V. Vishnu[1]

[1]Department of Sciences, Amrita School of Physical Sciences, Coimbatore, Amrita Vishwa Vidyapeetham, India

ks_rajani@cb.amrita.edu

Abstract

The future depends on sustainable energy resources; solar energy is Earth's most readily available energy source. To convert this massive energy source into electrical energy, we need to develop better photovoltaic technologies that are energy efficient and sustainable. Among the various photovoltaic technologies, the CZTS thin film solar cells are one of the safest absorber materials developed till now. Currently, the practical efficiency of CZTS needs to catch up to the theoretical efficiency and to reduce this gap, the materials used require fine-tuning. In this regard through this book chapter, we are trying to identify the motive behind the design and development of CZTS thin film solar cells and the future scope for quaternary chalcogenide solar cell.

Keyword

Cu_2ZnSnS_4, Thin Film Solar Cells, Chalcogenides, Solar Energy

Contents

1. Introduction

In the era of industrialization, nations aimed to achieve rapid economic and social development. The drawback of this type of development model is that it doesn't adequately consider environmental protection and sustainable development. Until the 20th century, the research and development community supported this type of development. The economic development was based on utilizing various energy resources for transportation, industrialization, and household applications. The energy resources available on Earth can be categorized into renewable and non-renewable energy sources, and the world depends on both energy sources for its development.

The various nonrenewable energy sources are primarily based on petroleum products and coal and are developed to their full potential. The drawback of these energy resources was that they were the primary sources of various pollutants which affect air, water, and soil. The alternatives to the nonrenewable energy sources are called renewable energy sources, which convert natural energy sources such as wind, solar, tidal, and geothermal energy into electrical and thermal energy. Hydroelectric and nuclear power plants are also part of renewable energy sources.

In the present sustainable development scenario, there is a massive shift of focus toward various renewable energy resources. The rapid development and commercialization of various renewable energy resources raises the question, are these so-called renewable energy resources sustainable? In the prima facie, these renewable energy resources are sustainable, but looking deeper into the various systems they could be more sustainable. For example, even though nuclear power plants are economical, there is always a threat of a nuclear disaster around them. Also, much effort is needed to recycle and store nuclear waste. The large-scale hydroelectric power plants sacrifice hectares of forest land to construct dams, are costly to develop, and are located in remote areas.

The problem with tidal, wind, and geothermal energy resources is that they are bound to geographical limitations. Thus, among the various renewable energy resources, only solar energy is available throughout the year without geographical constraints. Solar energy is converted into electrical energy using various photovoltaic technologies or thermal energy using various solar concentrators. The primary focus is on various photovoltaic technologies, as solar thermal power plants increase atmospheric temperature and affect the local ecosystem.

Thus, photovoltaic technology is the world's most reliable renewable energy source, which is economical and out of geographical constraints. If we again ask the above question in the context of solar cells, the answer should be a no.

The photovoltaic device can be classified into three generations (1st, $2^{nd,}$ and 3rd). The first-generation solar cells are made of Silicon (Si) wafers [1]. Even though Silicon is abundantly available in nature, producing these Si wafers is not energy efficient and has a very high material consumption. The second-generation solar cells are called thin film solar cells as they are made up of two-dimensional nanomaterials such as Gallium Arsenide (GaAs) [2], Cadmium Telluride (CdTe) [1], Copper Indium Gallium di Selenide (CIGS) [1], amorphous silicon [1] etc. Compared to the first-generation Si-based solar cells, these are easy to manufacture, but the issue is related to materials' availability and toxicity. Thus, the first two generations of solar cells cannot be considered sustainable. The third generation of solar cells focuses on using readily available non-toxic materials combined with less sophisticated fabrication techniques to make the solar cells sustainable.

The third-generation solar cells are based on nanotechnology and have Dye-Sensitized Solar Cells (DSSCs) [3], Quantum-Dot Sensitized Solar Cells (Q-DSSC) [4], Perovskite Solar Cells [5], and thin film solar cells [6]. This chapter focuses on the thin film solar cell based on Cu_2ZnSnS_4, an I-II-IV-VI type solar cell [6], according to the position of components in the periodic table.

1.1 I-II-IV-VI solar cells

The I-II-IV-VI type solar cell family consists of four primary members, namely Cu_2ZnSnS_4/Se_4[7,8] and Cu_2CdSnS_4/Se_4[9,10] and their alloys. This group of materials is considered a futuristic light absorber material for solar cells since they have a suitable band gap of 1.3-1.5 eV with a very high absorbance coefficient ($\sim 10^4$/cm) [11]. Another advantage of this group of materials is that they can be fabricated using a wide variety of methods such as chemical bath deposition (CBD) [11], Sputtering [12], SILAR [13], Solgel [14,15], spray pyrolysis [16], hydrothermal [17], solvothermal [18], pulsed laser deposition [19], electrodeposition [20], microwave synthesis [21] etc. Even though the I-II-IV-VI family consists of four members, the Copper Zinc Tin Sulphide (Cu_2ZnSnS_4/CZTS) is the one that scientists around the globe focus on. This is because CZTS is free of toxic materials like cadmium (Cd) and Selenium (Se), and all elements are abundant in nature.

Cu_2ZnSnS_4 and $Cu_2ZnSnSe$ exist in kesterite, stannite, and Primitive Mixed Cu-Au (PMCA) [22,23] crystal structures. Both kesterite and stannite structures are body-centered tetragonal structures with c=2a, while the PMCA is a primitive tetragonal structure with a=c [24]. The primary difference between these three structures is the difference in the atomic arrangement in the voids of the tetragonal structure (Figure 1). There are two alternating layers of Cu and Zn or Cu and Sn in the kesterite structure; in the stannite structure, the Zn and Sn atoms in the same plane switch positions in alternate layers. For the PMCA, there is no swapping of Zn and Sn in the alternate layers.

1.2 Cu_2ZnSnS_4 photovoltaics

The generation of electron-hole pairs is the basic working principle of every photovoltaic device, irrespective of its era. Generating electron-hole pairs depends on the material property. Thus, the material must have the following qualities to be used as the active material for solar cells:

- Effectively absorb photons having energy 1.0 eV-2.0 eV.

- Must be capable of generating electron-hole pairs.

- Efficient separation of electron-hole pairs.

- Formation of ohmic or quasi-ohmic contacts with metallic contacts for collecting and transporting electron-hole pairs.

The kesterite structure of CZTS has a suitable direct band gap of 1.4-1.5 eV coupled with a very high optical absorption coefficient, making it ideal for solar cell application. Compared to the other two structures, the kesterite structure is preferred owing to its excellent thermodynamical stability [25]. The Shockley Queisser limit for the CZTS solar cells is approximately 32.2%, while the highest efficiency reported is 12.6% [26]. This difference in theoretical and practical efficiencies allows further research into the I-II-IV-VI quaternary compounds for photovoltaic applications.

Materials Research Forum LLC
https://doi.org/10.21741/9781644903032-4

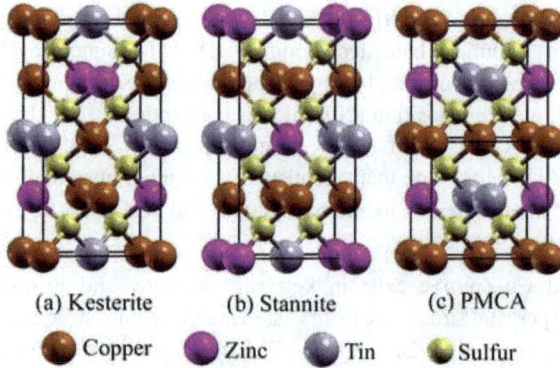

(a) Kesterite (b) Stannite (c) PMCA

Copper Zinc Tin Sulfur

Figure 1. Crystal structures of Cu_2ZnSnS_4 (Reproduced from Khare, Ankur, et al. "Calculation of the lattice dynamics and Raman spectra of copper zinc tin chalcogenides and comparison to experiments." Journal of Applied Physics 111.8 (2012): 083707. With permission from AIP publications).

1.3   Substrate and superstrate structures

Solar cells can be developed in two different structures (configurations): substrate and superstrate. The conventional CZTS solar cells were developed in the substrate configuration. In this configuration, either a conducting polymer or Mo-coated glass plate is used as the substrate material, which acts as the electrode to collect and transport the photogenerated holes and is the mechanical support to the device [27]. Mo has excellent stability and low resistivity and acts as the back contact. In addition, the presence of molybdenum disulfide (MoS_2) is beneficial for developing a better ohmic connection with CZTS. It assists in absorbing the solar spectrum in the visible region as it has an indirect band gap value of 1.3 eV. IN SOME CASES, the FTO/ITO-coated conductive glass plates are also utilized as back contact [28]. The absorber layer (CZTS) is then deposited on top of the back contact (Mo) is the active layer for the absorption of photons, and the thickness of CZTS will be in the range of 500 μm to 1mm [29]. An n-type buffer layer (normally CdS) of thickness less than 100 nm is then deposited on the CZTS to form the p-n heterojunction. Usually, the CdS interface is deposited via chemical bath deposition. The cell is then completed by depositing a layer of transparent conductive oxide (TCO). Usually, the TCO layer will be a bilayer of intrinsic zinc oxide (i-ZnO) and aluminum- doped zinc oxide (AZO) (Figure 2). An anti-reflection coating on the front contact enhances the photon capture by reducing the reflection losses and improving the charge collection [30].

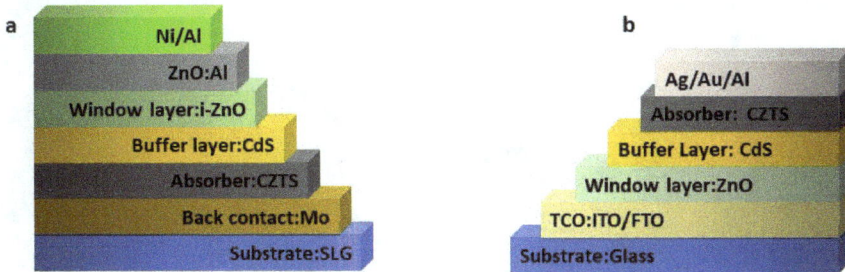

Figure 2. Structure of CZTS solar cell [21].

The advantage of this configuration is that both opaque and transparent materials can be used as the substrate. On the other hand, in superstrate configuration, the choice of material is limited to transparent conductive electrodes such as ITO/FTO. In addition to these, transparent flexible substrates can also be used. The superstrate structure is a reverse configuration of the substrate configuration, and the light is transmitted through the TCO into the active layer through the buffer layer. The cell configuration is completed by sputtering a metal contact to collect and transport photogenerated holes [31]. The working of the CZTS solar cell is illustrated in Figure 3.

While comparing the two structures of CZTS solar cells, the substrate configuration has outperformed the superstrate configuration concerning photovoltaic efficiency. Approximately a PCE of 13% is achieved in the substrate configuration, while the superstrate configuration shows less than 5% [28]. The drawback of substrate configuration is that the conductivity of the window layer (n-type) reduces due to aging since the CZTS is a coarse material. Low- temperature fabrication methods also degrade the quality of the buffer and window layer, resulting in recombination losses [31,33]. Therefore, the superstrate structure of CZTS solar cells is recommended as a solution for these issues due to the following advantages [28]:

- Lower cost of production.

- Facile processes.

- Easy and efficient light trapping.

- Easy to improve the design of back contact, and

- It is possible to develop tandem solar cells.

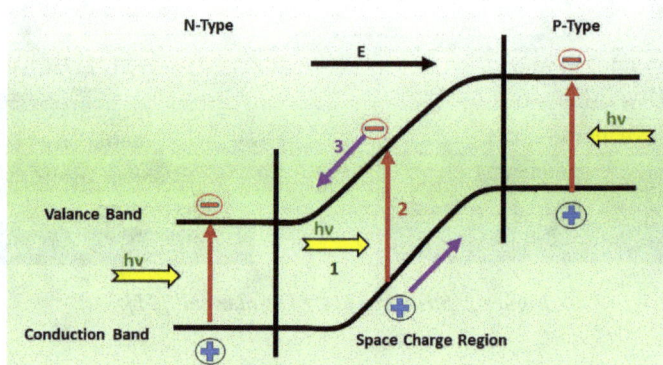

Figure 3: Working of CZTS solar cell [32].

2. CZTS photovoltaics fabrication

2.1 Spray pyrolysis

The chemical synthesis routes are known for their simplicity. Among the various chemical synthesis methods of thin films, spray pyrolysis is one of the most commonly employed methods due to its high reproducibility on a large scale. Another advantage of this method is that it does not require any vacuum conditions to deposit thin films. Fabrication of CZTS thin films using spray pyrolysis was first reported in 1996 by Nakayama and Ito [34]. They developed the stannite structure of CZTS with a bandgap of 1.46 eV. In 2007, Kamoun et al.

reported the kesterite structure of CZTS using spray pyrolysis by following a similar experimental procedure to that of Nakayama and Ito [35]. In 2012, Das et al. developed spray coated CZTS solar cell with an open circuit voltage (Voc) of 280mV and short circuit current density (Jsc) of $3.1mA/cm^2$ after employing an H_2S treatment at 350ºC [36]. A Photoconversion efficiency (PCE) of 7.1% was achieved by incorporating 2% Ag into the CZTS [37]. Lie et al. developed a CZTS solar cell of 6.73% efficiency by combining Mg with CZTS [38].

It is also found that the various parameters like pH of the precursor solution [39] solvent [40], substrate temperature [41], post-deposition annealing [42], spray duration [43], and concentration of Cu, Zn, and Sn etc. [44-47] influence the physical and chemical properties of the CZTS thin films and which in turn affects the performance of the CZTS solar cell developed through spray pyrolysis.

Third Generation Photovoltaic Technology   Materials Research Forum LLC
Materials Research Foundations 163 (2024) 92-117   https://doi.org/10.21741/9781644903032-4

2.2 Solgel method

The sol-gel-based techniques are simple and cost-effective. This method's advantage is that nanoparticles and thin films can be quickly developed with needed stoichiometry. The sol-gel- based spin coating is the most commonly used method for fabricating thin films. The advantage of this method is that the elemental precursors are mixed in a suitable solvent to form the sol, which then produces nanoparticles by polycondensation [48]. This method allows thin films with improved crystallinity to be developed at lower annealing temperatures by transferring the gel at reliable composition gel to thin films at room temperature [49].

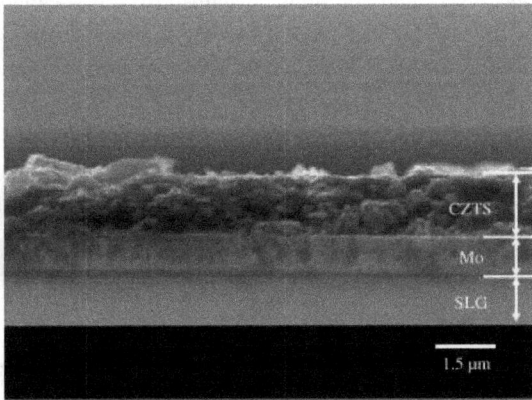

Figure 4: The layered structure of CZTS thin film was developed using spin coating [52].

The first sol-gel-based spin-coated CZTS thin film was reported by Tanaka et.al. in 2007[50]. The thin films were fabricated at 4000 rpm for 30 sec and dried at 400°C in the open atmosphere. Finally, the fabricated films were annealed at 500°C in an N_2+ 5%H_2S atmosphere for 1 hour. The effect of pre-annealing on spin-coated CZTS thin films resulted in a red shift of the energy gap [51]. In 2009, a spin-coated CZTS thin film solar cell with a PCE of 1.01% efficiency (figure 4) is reported by Tanaka et. al [52]. The superstrate structured Au/CZTS/ZnS/ZnO NR/ITO solar cell developed using sol-gel-based spin coating is the current champion cell in superstrate geometry with a photoconversion efficiency of 3.63% [53]. The spin-coated CZTSSe thin film solar cell with additional MgF_2 anti-reflection coating produced a high % photoconversion efficiency of 10.1% in 2011 [54]. Currently, the champion cell of CZTS is developed using the spin coating and has a PCE of 12.1% [55].

2.3 Hydrothermal, solvothermal, and microwave-assisted synthesis

The hydrothermal method and its derivatives (solvothermal, microwave-assisted) commonly utilize chemical routes for nanoparticle and thin-film synthesis [56-59]. Hydrothermal originated from geology, which refers to using an aqueous reaction system heated in a closed vessel called autoclave to create high temperature and pressure. In this process, the reactants dissolve and recrystallize in the high temperature and pressure created in the autoclave vessel [60,61]. In the conventional hydrothermal method, the heat is supplied using a hot air oven; the time and temperature of the reaction, the concentration of reactants, the pH of the solution, etc., influence the physical and hence the chemical properties of the nanocrystals and thin films developed [61]. It is called the solvothermal method if an organic solvent is used instead of water [58]. In modern days, the conventional hot air oven is replaced by the microwave oven to speed up the reactions and to save energy [62].

The first report of the fabrication of CZTS using the hydrothermal method was reported in 2011. Both the stannite and wurtzite structures of CZTS were reported separately in that year [63,64]. The hydrothermal method of CZTS nanoparticle synthesis is illustrated in Figure 5. The development of CZTS through the hydrothermal method has been reported many times, but the number of solar cells reported is significantly less. Patil et al., fabricated a CZTS thin film solar cell and Photoelectrochemical (PEC) cell with a PCE of 4.82% and 3.21%, respectively, in 2021[66,67]. The CZTS can also be utilized as a counter electrode for dye- sensitized solar cells (DSSC), and a PCE of 2.65% was reported in 2019 [68].

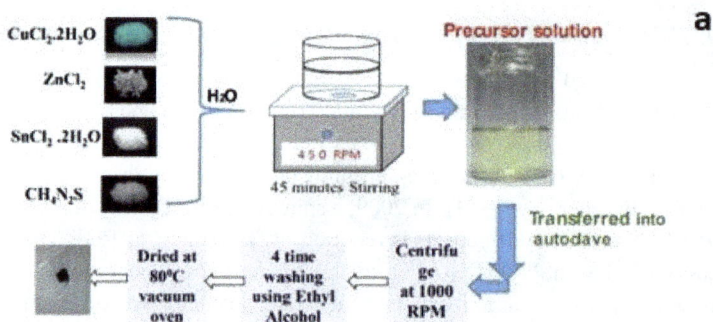

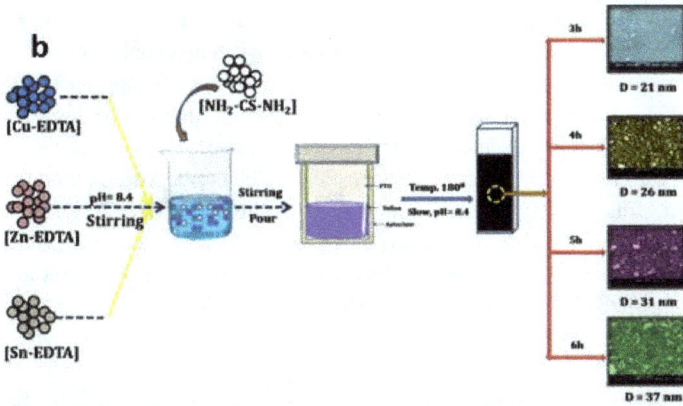

Figure 5: The hydrothermal method a. CZTS nanoparticle b. thin films [65,67].

2.4　SILAR method and chemical bath deposition (CBD)

The Successive Ion Layer Adsorption and Reaction (SILAR) method and Chemical Bath Deposition (CBD) are two of the most simple and common chemical synthesis routes for the deposition of chalcogenide thin films [69,70]. The advantage of the SILAR method is that no sophisticated instrument is needed, and the thin film deposition can be done at room temperature. The parameters like concentration of precursor solution, pH, number of deposition cycles, adsorption, reaction and rinsing time, complexing agent etc. affect the quality of the thin films deposited through the SILAR method [69]. In CBD, the deposition time, the temperature of deposition, the pH of the precursor solution, the concentration of the solution etc. affect the physical and chemical properties of thin films [70].

Shinde et al., developed the CZTS absorber layer developed through the SILAR method and reported a PCE of 0.12%. The kesterite-structured thin films annealed at 673°C had a homogeneous and dense surface property with a direct band gap of 1.55 eV [71]. The CdS/CZTS superstrate solar cell developed by Kaza et al [72] had an efficiency of 2.45%. The electrical properties of CdS/CZTS heterojunction were studied by Krishnan et. al., and reported junction had a resistivity of 1.51×10^2 Ω cm, with a carrier concentration of $1.28 \times 10^{17} cm^{-3}$ [73]. Most of the CZTS thin films developed through the SILAR method are utilized to develop PEC cells [74-76].

2.5　Electrodeposition

Electrodeposition is another exciting method utilized for the production of semiconductor thin films for research applications as well as for industrial applications. Photovoltaic

Third Generation Photovoltaic Technology   Materials Research Forum LLC
Materials Research Foundations 163 (2024) 92-117   https://doi.org/10.21741/9781644903032-4

materials such as Cadmium Telluride, copper indium sulfide, copper indium gallium di-selenide, CZTS etc. are fabricated through this method. Sequential electroplating or one-step electrodeposition and simultaneous deposition are the two approaches to thin film deposition [21]. The one-step electrodeposition is commonly used to deposit binary thin films like the CdTe. In this method, the metal precursors are sequentially stacked in accordance with the stoichiometry and deposited via electrodeposition, and then undergo reactive annealing in the preferred atmosphere (sulfur/selenide) [77]. In the second method, the three metallic precursors are simultaneously electrodeposited and then undergo the annealing treatment at sulfur or selenide atmosphere. Berrut et al. developed an electrodeposited CZTS thin film solar cell with an efficiency of 3.5% [78].

2.6   Screen printing

Screen printing is one of the less sophisticated, economical methods for thin film fabrication in the modern days (Figure 6). This technique can fabricate various types of thin films, including oxides, sulfides, selenides, tellurides, and even carbon-based thin films can be prepared [79]. These printed thin films find applications in energy conversion and storage, sensors, flexible electronics, etc. the most significant advantage of the screen-printing technique is that it enables roll-to-roll (R2R) manufacturing of various components, including flexible solar cells and batteries [80]. Zhou et al., developed a thin film CZTS solar cell with a PCE of 0.49% utilizing screen printing [81]. Wang et al. studied the effect of sintering temperature on the quality of CZTS thin films deposited via screen printing [82].

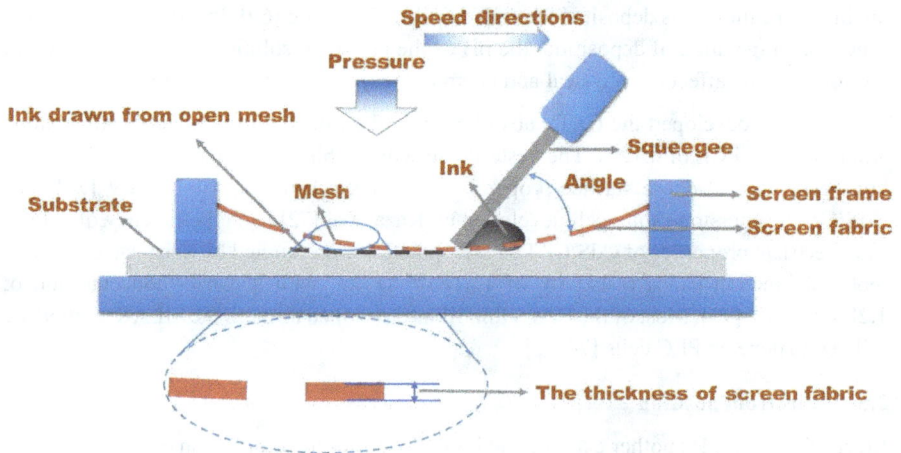

Figure 6: Screen Printing Schematic diagram [80]

Qinmiao et al., developed an FTO/TiO$_2$/In$_2$S$_3$/CZTS/Carbon solar cell in the superstrate configuration and reported an efficiency of 0.6% with a fill factor of 0.27. The CZTS layer was developed by combining ball milling and screen-printing techniques, and all other layers were fabricated using spray coating [83]. The doctor-blade method can be treated as a manual thin film printing method which is less complicated than screen printing. The drawback of the doctor blade method is that it needs precision while coating the material. Compared with the screen-printing method, the large-area coating is challenging through the doctor's blade technique [30]. Chen et.al., developed an FTO/TiO$_2$/In$_2$S$_3$/CZTS/Mo in superstrate configuration with 0.55% PCE using the doctor blade method [84].

Table 1: Performance of CZTS solar cell with various deposition methods.

S.No	Device Architecture	Absorber layer	Deposition method	PCE (%)	Ref
	FTO/TiO$_2$/In$_2$S$_3$/CZTS/Mo	CZTS	Ball milling/doctor blade	0.55	[84]
	FTO/TiO$_2$/In$_2$S$_3$/CZTS/Carbon	CZTS	Screen Printing	0.60	[83]
	polyimide/Mo/Cu2ZnSnS4/CdS/ZnO:Al/Al-grid	CZTS	Screen Printing	0.49	[81]
	FTO/TiO2/In2S3/Cu2ZnSnS4/graphite	CZTS	Electrodeposition	3.5	[78]
	FTO/CdS/CZTS/Ag	CZTS	SILAR	2.45	[72]
	Mo/CZTS/CdS/Ag	CZTS	SILAR	1.02	
	FTO/CZTS/Eu(NO3)3/graphite	CZTS	Hydrothermal	4.87	[66]
	Al/ZnO:Al/CdS/CZTS/Mo/SLG	CZTS	Spin coating	1.01	[52]
	ITO/ZnO (NR's)/ZnS/CZTS/Au	CZTS	Spin Coating	3.63	[53]
0	SLG/Mo/CZTSSe/CdS/ZnS/ITO	CZTSSe	Spin Coating	10.1	[54]
1	SLG/Mo/Mo(S,Se)$_2$/CZTSSe/CdS/ZnS/ITO	CZTSSe	Spin Coating	12.6	[55]
2	SLG/Mo/CZTS/CdS/ITO	CZTS	Spray Coating	-----	[36]
3	Glass/Mo/ACZTS/CdS/Al	AgCZTS	Spray Coating	7.1	[37]
4	Glass/Mo/Cu2MgxZn1-xSnS4/CdS/ITO	Cu$_2$Mg$_x$Zn$_{1-x}$SnS$_4$	Spray Coating	6.73	[38]
5	SLG/ITO/ZnO-NR/CdS/CZTS/Ag	CZTS	One-step thermal evaporation	2.82	[85]
6	Glass/Mo/CZTS/CdS/i-ZnO/AZO/Ag	CZTS	Microwave Assisted Synthesis	0.25	[86]
7	FTO/TiO$_2$CdS/CZTS/Au	CZTS	Hydrothermal	1.45	[87]
8	ITO/ZnO/CZTS/Carbon	CZTS	Microwave Assisted synthesis	0.85	[88]
9	Willow glass/Mo/CZTS/CdS/ZnS/ITO	CZTS	Magneton Sputtering	3.08	[89]
0	SS/Cr/Mo/ZnO/CZTS/CdS/AZO	CZTS	Co-Sputtering	3.5	[90]

3. Absorber material

The active or absorber material is the key layer of a solar cell. The absorber layer's structural, morphological, optical, and electrical properties have a significant role in the performance of any photovoltaic device. The optical properties are directly dependent on the absorber material's band gap and absorption coefficient. At the same time, the charge transfer is related to the electronic structure and band positioning. The material's morphology enhances the photon absorption of the material by reducing the loss by reflection. A lattice match between the adjacent layers is required for perfect alignment without any defects, and this avoids the formation of charge recombination centers.

The quaternary I-II-IV-VI compound, such as the $Cu_2ZnSn(S_4/Se_4)$ and $Cu_2CdSn(S_4/Se_4)$ are developed by substituting Indium (In) in $CuInS_2$ with equal amounts of Zn/Cd and Sn. Charge neutrality is maintained even after substituting with these materials [91]. Compared with binary and ternary semiconducting materials, quaternary semiconductors give flexibility to the researchers in selecting the materials and tuning the properties. The main drawback of quaternary semiconductors such as CZTS/Se is the formation of various defects in the crystal structure, which may alter the absorber material's structural, optical, electrical, and photovoltaic properties. The commonly found defects in CZTS/Se are the following [21]:

- Vacancies of Cu, Zn, Sn and S.
- Anti-sites A_B (A replaces B; Cu_{Zn}, Sn_{Zn}, Sn_{Cu} Cu_{Sn}, and Zn_{Sn}).
- Interstitial defects related to Cu, Zn, and Sn.
- Formation of Complexes and Clusters.

In addition, there is a possibility of forming a mixed Kesterite/Stannite phase due to the interchanging of Cu and Zn positions since both have similar ionic radii ($Cu^{2+} \approx$ pm and $Zn^{2+} \approx 134$ pm) [92]. Further, the anti-site defects related to Zn in CZTS/Se result in low open circuit voltage (V_{OC}) due to the formation of band-tailed states, limiting the photoconversion efficiency [93,94]. Thus, researchers examined the $Cu_2CdSn(S_4/Se_4)$ as the absorber material in solar cells. The Cd^{2+} has a larger ionic ($Cd^{2+} \approx 151$ pm) radius than that of Cu^{2+} and Zn^{2+}, reducing the formation of mixed phases in CZTS/Se [92]. Thus, in addition to various fabrication techniques, various quaternary absorber materials can also be utilized for photovoltaic applications. Among the various quaternary absorber materials, the Cu_2CdSnS_4 has achieved a PCE of 10.14% [92]. Like the Cu_2ZnSnS_4 and Cu_2CdSnS_4, several quaternary chalcogenides can be utilized for solar cell applications (Table 2).

In addition to the formation of defects, the quaternary compounds such as Cu_2ZnSnS_4/Se_4, Cu_2CdSnS_4, Cu_2FeSnS_4, Cu_2CoSnS_4, Cu_2SrSnS_4, etc., are prone to secondary phase formations due to the presence of four or more elements. In Cu_2ZnSnS_4, the development of secondary phases like CuS, Cu_2S, SnS, ZnS, Cu_2SnS_3, etc., are reported many times [101]. The formation of the pure CZTS phase is complicated and has narrow stability [102]. It is clear from the chemical potential diagram that the boundaries of CZTS are various secondary phases, and the region of formation of CZTS is narrow (figure 7).

The commonly observed secondary phases in CZTS are ZnS, Cu_2S, SnS_2, and Cu_2SnS_3. ZnS is usually a wide band gap (3.54 eV) material, which decreases the active area for solar absorption, electron-hole generation, and current collection. Cu_2S is a very good electrical conductor with a band gap of 1.21 eV and tends to short-circuit CZTS solar cells. SnS_2 is an n- type material with a bandgap value of 2.2 eV and tends to form secondary diodes inside the absorber material and can cause high charge recombination. The p-type Cu_2SnS_3 has a band gap similar to CZTSSe. Still, it is less efficient in generating photogenerated charge carriers, and reducing the unwanted secondary phases is one of the most challenging parts while developing CZTS solar cells [101].

Table 2: Various quaternary absorber materials and their cell performance.

S. No	Device Architecture	Absorber layer	Deposition method	PCE (%)	Ref
1	SLG/Mo/CCTS/CdS/ITO/Ag	Cu_2CdSnS_4	Spin Coating	10.14	[92]
2	SLG/Mo/Mo(S,Se)$_2$/CZTSSe/CdS/ZnS/ITO	Cu_2ZnSnS_4/Se$_4$	Spin Coating	12.6	[55]
3	Glass/Mo/Cu$_2$Mg$_x$Zn$_{1-x}$SnS$_4$/CdS/ITO	$Cu_2Mg_xZn_{1-x}SnS_4$	Spray Coating	6.73	[38]
4	Glass/Mo/ACZTS/CdS/Al	AgCZTS	Spray Coating	7.1	[37]
5	ITO/Cu@NiO/CFTS/Bi$_2$S$_3$/ZnO/Al	Cu_2FeSnS_4	SILAR	2.9	[95]
6	FTO/CBTS/CdS/i-ZnO/ITO/Ag	Cu_2BaSnS_4	Co-Sputtering	1.21	[96]
7	Al/n-Si/Cu$_2$CoSnS$_4$	Cu_2CoSnS_4	Spray coating	6.71	[97]
8	Ag/n-Si/CNSS/Au	Cu_2NiSnS_4	Spray pyrolysis	11.34	[98]
9	Mo/AZTS(QD)/CdS/ZnO/Al	Ag_2ZnSnS_4	Solvothermal	6.28	[99]
10	Mo/MoS$_2$/CSTS/CdS/ZnO/ITO	Cu_2SrSnS_4	Co-Sputtering	0.54	[100]

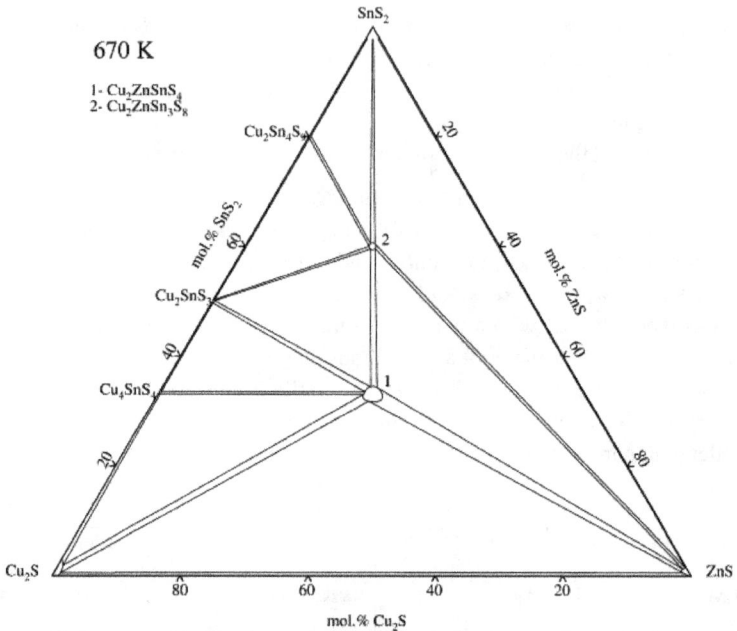

Figure 7: The pseudo ternary phase diagram of CZTS [102].

4. Future scope

The quaternary chalcogenides are the future candidates for better photovoltaic devices. These materials can be utilized in various solar cells, including thin film solar cells, DSSCs, photoelectrochemical cells, tandem solar cells, and hybrid solar cells. The materials also find applications in energy storage devices [104]. Among these quaternary chalcogenides, the CZTS is vital since the material is earth-abundant, cost-effective, and non-toxic. Another advantage of the material is that more than one method can be employed for thin film fabrication, including sophisticated vacuum-based and simple solution-based methods.

The structural properties determine the optoelectronic functioning of the CZTS. A fine-tuning of the CZTS structure related is required to restrict the presence of secondary phases, defects, impurities, vacancies, dislocations, etc. the fine-tuning of the optical band gap can be done by doping the absorber material with a suitable material like Na, Ag, and also, by incorporating Cadmium and Selenide to replace Copper and Sulfur sites respectively [105].

An Appropriate back contact, window layer, buffer layer, and TCO are also essential for fine-tuning the photovoltaic capabilities of CZTS.

Acknowledgment

The authors sincerely acknowledge Dr. A. Pandikumar, Senior Scientist, Electro Organic and Materials Electrochemistry Division, CSIR - Central Electrochemical Research Institute, Karaikudi, for allowing us to write this book chapter.

References

[1] Imamzai, Mohammadnoor, et al. "A review on comparison between traditional silicon solar cells and thin-film CdTe solar cells." Proceedings of National Graduate Conference (Nat-Grad. 2012.

[2] Radziemska, Ewa. "Thermal performance of Si and GaAs based solar cells and modules: a review." Progress in Energy and Combustion Science 29.5 (2003): 407-424. https://doi.org/10.1016/S0360-1285(03)00032-7

[3] Raguram, T., and K. S. Rajni. "Synthesis and characterisation of Cu-doped TiO2 nanoparticles for DSSC and photocatalytic applications." International Journal of Hydrogen Energy 47.7 (2022): 4674-4689. https://doi.org/10.1016/j.ijhydene.2021.11.113

[4] Nozik, Arthur J. "Quantum dot solar cells." Physica E: Low-dimensional Systems and Nanostructures 14.1-2 (2002): 115-120. https://doi.org/10.1016/S1386-9477(02)00374-0

[5] Kim, Jin Young, et al. "High-efficiency perovskite solar cells." Chemical Reviews 120.15 (2020): 7867-7918. https://doi.org/10.1021/acs.chemrev.0c00107

[6] Peksu, Elif, and Hakan Karaagac. "Characterization of Cu2ZnSnS4 thin films deposited by one- step thermal evaporation for a third generation solar cell." Journal of Alloys and Compounds 862 (2021): 158503. https://doi.org/10.1016/j.jallcom.2020.158503

[7] Jhuma, Farjana Akter, Marshia Zaman Shaily, and Mohammad Junaebur Rashid. "Towards high- efficiency CZTS solar cell through buffer layer optimization." Materials for Renewable and Sustainable Energy 8 (2019): 1-7. https://doi.org/10.1007/s40243-019-0144-1

[8] Giraldo, Sergio, et al. "Large efficiency improvement in Cu2ZnSnSe4 solar cells by introducing a superficial Ge nanolayer." Advanced Energy Materials 5.21 (2015): 1501070. https://doi.org/10.1002/aenm.201501070

[9] Fan, Ping, et al. "Over 10% Efficient Cu2CdSnS4 Solar Cells Fabricated from Optimized Sulfurization." Advanced Functional Materials 32.45 (2022): 2207470. https://doi.org/10.1002/adfm.202207470

[10] Basu, Ranita, et al. "Improving the Thermoelectric Performance of Tetrahedrally Bonded Quaternary Selenide Cu 2 CdSnSe 4 Using CdSe Precipitates." Journal of Electronic Materials 48 (2019): 2120-2130. https://doi.org/10.1007/s11664-019-07012-0

[11] Macías-Cabrera, C. A., et al. "Synthesis of CZTS thin films from binary precursors stacking by chemical bath deposition for solar cell applications." Materials Today: Proceedings 46 (2021): 3109-3113. https://doi.org/10.1016/j.matpr.2021.02.624

[12] Sharmin, Afrina, et al. "Sputtered single-phase kesterite Cu2ZnSnS4 (CZTS) thin film for photovoltaic applications: Post annealing parameter optimization and property analysis." AIP Advances 10.1 (2020): 015230. https://doi.org/10.1063/1.5129202

[13] Krishnan, Ambily, et al. "Towards phase pure CZTS thin films by SILAR method with augmented Zn adsorption for photovoltaic applications." Materials for Renewable and Sustainable Energy 8 (2019): 1-8. https://doi.org/10.1007/s40243-019-0152-1

[14] Ashfaq, Arslan, et al. "A two step technique to remove the secondary phases in CZTS thin films grown by sol-gel method." Ceramics International 45.8 (2019): 10876-10881. https://doi.org/10.1016/j.ceramint.2019.02.165

[15] Ahmoum, H., et al. "Impact of preheating environment on microstructural and optoelectronic properties of Cu2ZnSnS4 (CZTS) thin films deposited by spin-coating." Superlattices and Microstructures 140 (2020): 106452. https://doi.org/10.1016/j.spmi.2020.106452

[16] Prabeesh, P., et al. "Influence of thiourea in the precursor solution on the structural, optical and electrical properties of CZTS thin films deposited via spray coating technique." Journal of Materials Science: Materials in Electronics 32 (2021): 4146-4156. https://doi.org/10.1007/s10854-020-05156-y

[17] Sawant, Jitendra P., and Rohidas B. Kale. "CZTS counter electrode in dye-sensitized solar cell: enhancement in photo conversion efficiency with morphology of TiO 2 nanostructured thin films." Journal of Solid State Electrochemistry 24 (2020): 461-472. https://doi.org/10.1007/s10008-019-04452-w

[18] Mkawi, E. M., et al. "Fabricating chalcogenide Cu2ZnSnS4 (CZTS) nanoparticles via solvothermal synthesis: Effect of the sulfur source on the properties." Ceramics International 46.16 (2020): 24916-24922. https://doi.org/10.1016/j.ceramint.2020.06.276

[19] Elhmaidi, Z. O., et al. "In-situ tuning of the zinc content of pulsed-laser-deposited CZTS films and its effect on the photoconversion efficiency of p-CZTS/n-Si heterojunction photovoltaic devices." Applied Surface Science 507 (2020): 145003. https://doi.org/10.1016/j.apsusc.2019.145003

[20] Borate, Haribhau, et al. "Single-step electrochemical deposition of CZTS thin films with enhanced photoactivity." ES Materials & Manufacturing 11 (2020): 30-39.

[21] Yussuf, Sodiq Tolulope, et al. "PHOTOVOLTAIC EFFICIENCIES OF MICROWAVE AND Cu2ZnSnS4 (CZTS) SUPERSTRATE SOLAR CELLS." Materials Today Sustainability (2022): 100287. https://doi.org/10.1016/j.mtsust.2022.100287

[22] Kumar, Mukesh, and Clas Persson. "Cu2ZnSnS4 and Cu2ZnSnSe4 as potential earth-abundant thin-film absorber materials: a density functional theory study." Int. J. Theor. Appl. Sci 5.1 (2013): 1-8. https://doi.org/10.1063/1.4812448

[23] Kattan, Nessrin, et al. "Crystal structure and defects visualization of Cu2ZnSnS4 nanoparticles employing transmission electron microscopy and electron diffraction." Applied Materials Today (2015): 52-59. https://doi.org/10.1016/j.apmt.2015.08.004

[24] Khare, Ankur, et al. "Calculation of the lattice dynamics and Raman spectra of copper zinc tin chalcogenides and comparison to experiments." Journal of Applied Physics 111.8 (2012): 083707. https://doi.org/10.1063/1.4704191

[25] Schorr, S. J. T. S. F. "Structural aspects of adamantine like multinary chalcogenides." Thin Solid Films 515.15 (2007): 5985-5991. https://doi.org/10.1016/j.tsf.2006.12.100

[26] Khelfane, Amar, et al. "Composition dependence of the optical band gap and the secondary phases via zinc content in CZTS material." Inorganic Chemistry Communications (2023): 110639. https://doi.org/10.1016/j.inoche.2023.110639

[27] Rakitin, Vladimir V., and Gennady F. Novikov. "Third-generation solar cells based on quaternary copper compounds with the kesterite-type structure." Russian Chemical Reviews 86.2 (2017): 99. https://doi.org/10.1070/RCR4633

[28] Satale, Vinayak Vitthal, and S. Venkataprasad Bhat. "Superstrate type CZTS solar cell with all solution processed functional layers at low temperature." Solar Energy 208 (2020): 220-226. https://doi.org/10.1016/j.solener.2020.07.055

[29] Rand, Barry P., et al. "Solar cells utilizing small molecular weight organic semiconductors." Progress in Photovoltaics: Research and Applications 15.8 (2007): 659-676. https://doi.org/10.1002/pip.788

[30] Ghediya, Prashant R., and Tapas K. Chaudhuri. "Doctor-blade printing of Cu 2 ZnSnS 4 films from microwave-processed ink." Journal of Materials Science: Materials in Electronics 26 (2015): 1908- 1912. https://doi.org/10.1007/s10854-014-2628-1

[31] Gong, Yuancai, et al. "Elemental de-mixing-induced epitaxial kesterite/CdS interface enabling 13%-efficiency kesterite solar cells." Nature Energy 7.10 (2022): 966-977. https://doi.org/10.1038/s41560-022-01132-4

[32] Karzazi, Yasser, and Imane Arbouch. "Inorganic photovoltaic cells: Operating principles, technologies and efficiencies-Review." J. Mater. Environ. Sci 5 (2014): 1505-1515.

[33] Das, Sandip. Growth, fabrication and characterization of Cu 2 ZnSn (S x Se 1-x) 4 photovoltaic absorber and thin-film heterojunction solar cells. Diss. University of South Carolina, 2014.S

[34] Nakayama, Norio, and Kentaro Ito. "Sprayed films of stannite Cu2ZnSnS4." Applied Surface Science 92 (1996): 171-175. https://doi.org/10.1016/0169-4332(95)00225-1

[35] Kamoun, N., H. Bouzouita, and B. J. T. S. F. Rezig. "Fabrication and characterization of Cu2ZnSnS4 thin films deposited by spray pyrolysis technique." Thin Solid Films 515.15 (2007): 5949-5952. https://doi.org/10.1016/j.tsf.2006.12.144

[36] Das, Sandip, et al. "Deposition and characterization of low-cost spray pyrolyzed Cu2ZnSnS4 (CZTS) thin-films for large-area high-efficiency heterojunction solar cells." ECS Transactions 45.7 (2012): 153. https://doi.org/10.1149/1.3701535

[37] Nguyen, Thi Hiep, et al. "Structural and solar cell properties of a Ag-containing Cu2ZnSnS4 thin film derived from spray pyrolysis." ACS applied materials & interfaces 10.6 (2018): 5455-5463. https://doi.org/10.1021/acsami.7b14929

[38] Lie, Stener, et al. "Improving carrier-transport properties of CZTS by Mg incorporation with spray pyrolysis." ACS applied materials & interfaces 11.29 (2019): 25824-25832. https://doi.org/10.1021/acsami.9b05244

[39] Kumar, YB Kishore, et al. "Effect of starting-solution pH on the growth of Cu2ZnSnS4 thin films deposited by spray pyrolysis." Physica status solidi (a) 206.7 (2009): 1525-1530. https://doi.org/10.1002/pssa.200824424

[40] Gunavathy, K. V., et al. "Effect of solvent on the characteristic properties of nebulizer spray pyrolyzed Cu2ZnSnS4 absorber thin films for photovoltaic application." Thin Solid Films 697 (2020): 137841. https://doi.org/10.1016/j.tsf.2020.137841

[41] Bakr, Nabeel A., Ziad T. Khodair, and S. M. Hassan. "Effect of substrate temperature on structural and optical properties of Cu2ZnSnS4 (CZTS) films prepared by chemical spray pyrolysis method." Research Journal of Chemical Sciences ISSN 2231 (2015): 606X.

[42] Seboui, Zeineb, et al. "Effect of annealing process on the properties of Cu2ZnSnS4 thin films." Superlattices and Microstructures 75 (2014): 586-592. https://doi.org/10.1016/j.spmi.2014.07.025

[43] Kamoun, N., H. Bouzouita, and B. J. T. S. F. Rezig. "Fabrication and characterization of Cu2ZnSnS4 thin films deposited by spray pyrolysis technique." Thin Solid Films 515.15 (2007): 5949-5952. https://doi.org/10.1016/j.tsf.2006.12.144

[44] Rajeshmon, V. G., et al. "Effect of copper concentration and spray rate on the properties Cu2ZnSnS4 thin films deposited using spray pyrolysis." Journal of Analytical and Applied Pyrolysis 110 (2014): 448-454. https://doi.org/10.1016/j.jaap.2014.10.014

[45] Boutebakh, F. Z., et al. "Zinc molarity effect on Cu 2 ZnSnS 4 thin film properties prepared by spray pyrolysis." Journal of Materials Science: Materials in Electronics 29 (2018): 4089-4095. https://doi.org/10.1007/s10854-017-8353-9

[46] Thiruvenkadam, S., et al. "Effect of Zn/Sn molar ratio on the microstructural and optical properties of Cu2Zn1-xSnxS4 thin films prepared by spray pyrolysis technique." Physica B: Condensed Matter 533 (2018): 22-27. https://doi.org/10.1016/j.physb.2017.12.065

[47] Sampath, M., et al. "Structural, optical and photocatalytic properties of spray deposited Cu2ZnSnS4 thin films with various S/(Cu+ Zn+ Sn) ratio." Materials Science in Semiconductor Processing 87 (2018): 54-64. https://doi.org/10.1016/j.mssp.2018.07.001

[48] Agawane, G. L., et al. "Synthesis of simple, low cost and benign sol-gel Cu 2 ZnSnS 4 thin films: influence of different annealing atmospheres." Journal of materials

science: Materials in electronics 26 (2015): 1900-1907.
https://doi.org/10.1007/s10854-014-2627-2

[49] Song, Xiangbo, et al. "A review on development prospect of CZTS based thin film solar cells." International Journal of Photoenergy 2014 (2014). https://doi.org/10.1155/2014/613173

[50] Tanaka, Kunihiko, Noriko Moritake, and Hisao Uchiki. "Preparation of Cu2ZnSnS4 thin films by sulfurizing sol-gel deposited precursors." Solar Energy Materials and Solar Cells 91.13 (2007): 1199-1201. https://doi.org/10.1016/j.solmat.2007.04.012

[51] Tanaka, Kunihiko, et al. "Pre-annealing of precursors of Cu2ZnSnS4 thin films prepared by sol- gel sulfurizing method." Japanese journal of applied physics 47.1S (2008): 598. https://doi.org/10.1143/JJAP.47.598

[52] Tanaka, Kunihiko, et al. "Cu2ZnSnS4 thin film solar cells prepared by non-vacuum processing." Solar Energy Materials and Solar Cells 93.5 (2009): 583-587. https://doi.org/10.1016/j.solmat.2008.12.009

[53] Ghosh, Anima, Rajalingam Thangavel, and Arunava Gupta. "Solution-processed Cd free kesterite Cu2ZnSnS4 thin film solar cells with vertically aligned ZnO nanorod arrays." Journal of alloys and Compounds 694 (2017): 394-400. https://doi.org/10.1016/j.jallcom.2016.09.325

[54] Barkhouse, D. Aaron R., et al. "Device characteristics of a 10.1% hydrazine-processed Cu2ZnSn (Se, S) 4 solar cell." Progress in Photovoltaics: Research and Applications 20.1 (2012): 6-11. https://doi.org/10.1002/pip.1160

[55] Wang, Wei, et al. "Device characteristics of CZTSSe thin-film solar cells with 12.6% efficiency." Advanced energy materials 4.7 (2014): 1301465. https://doi.org/10.1002/aenm.201301465

[56] Camara, Sekou Mariama, Lingling Wang, and Xintong Zhang. "Easy hydrothermal preparation of Cu2ZnSnS4 (CZTS) nanoparticles for solar cell application." Nanotechnology 24.49 (2013): 495401. https://doi.org/10.1088/0957-4484/24/49/495401

[57] Patil, Satish S., et al. "Facile designing and assessment of photovoltaic performance of hydrothermally grown kesterite Cu2ZnSnS4 thin films: influence of deposition time." Solar Energy 201 (2020): 102-115. https://doi.org/10.1016/j.solener.2020.02.089

[58] Wei, Aixiang, et al. "Solvothermal synthesis of Cu2ZnSnS4 nanocrystalline thin films for application of solar cells." International Journal of Hydrogen Energy 40.1 (2015): 797-805. https://doi.org/10.1016/j.ijhydene.2014.09.047

[59] Madiraju, Venkata Alekhya, et al. "CZTS synthesis in aqueous media by microwave irradiation." Journal of Materials Science: Materials in Electronics 27 (2016): 3152-3157. https://doi.org/10.1007/s10854-015-4137-2

[60] Morey, George W., and Paul Niggli. "The hydrothermal formation of silicates, a review." Journal of the American Chemical Society 35.9 (1913): 1086-1130. https://doi.org/10.1021/ja02198a600

[61] Yoshimura, Masahiro, and Hiroyuki Suda. "Hydrothermal processing of hydroxyapatite: past, present, and future." Hydroxyapatite and related materials. CRC Press, 2017. 45-72. https://doi.org/10.1201/9780203751367-3

[62] Chin, Clare Davis-Wheeler, LaRico J. Treadwell, and John B. Wiley. "Microwave synthetic routes for shape-controlled catalyst nanoparticles and nanocomposites." Molecules 26.12 (2021): 3647. https://doi.org/10.3390/molecules26123647

[63] Wang, Chunrui, et al. "Synthesis of Cu2ZnSnS4 nanocrystallines by a hydrothermal route." Japanese Journal of Applied Physics 50.6R (2011): 065003. https://doi.org/10.1143/JJAP.50.065003

[64] Li, Mei, et al. "Synthesis of pure metastable wurtzite CZTS nanocrystals by facile one-pot method." The Journal of Physical Chemistry C 116.50 (2012): 26507-26516. https://doi.org/10.1021/jp307346k

[65] Das, S., et al. "Synthesis of quaternary chalcogenide CZTS nanoparticles by a hydrothermal route." IOP Conference Series: Materials Science and Engineering. Vol. 338. No. 1. IOP Publishing, 2018. https://doi.org/10.1088/1757-899X/338/1/012062

[66] Patil, Satish S., et al. "Optoelectronic and photovoltaic properties of the Cu2ZnSnS4 photocathode by a temperature-dependent facile hydrothermal route." Industrial & Engineering Chemistry Research 60.21 (2021): 7816-7825. https://doi.org/10.1021/acs.iecr.1c00801

[67] Patil, Satish S., et al. "Facile designing and assessment of photovoltaic performance of hydrothermally grown kesterite Cu2ZnSnS4 thin films: influence of deposition time." Solar Energy 201 (2020): 102-115. https://doi.org/10.1016/j.solener.2020.02.089

[68] Sawant, Jitendra P., and Rohidas B. Kale. "CZTS counter electrode in dye-sensitized solar cell: enhancement in photo conversion efficiency with morphology of TiO 2

nanostructured thin films." Journal of Solid State Electrochemistry 24 (2020): 461-472. https://doi.org/10.1007/s10008-019-04452-w

[69] Pathan, H. M., and C. D. Lokhande. "Deposition of metal chalcogenide thin films by successive ionic layer adsorption and reaction (SILAR) method." Bulletin of Materials Science 27 (2004): 85-111. https://doi.org/10.1007/BF02708491

[70] Mane, R. S., and C. D. Lokhande. "Chemical deposition method for metal chalcogenide thin films." Materials Chemistry and physics 65.1 (2000): 1-31. https://doi.org/10.1016/S0254-0584(00)00217-0

[71] Shinde, N. M., et al. "Room temperature novel chemical synthesis of Cu2ZnSnS4 (CZTS) absorbing layer for photovoltaic application." Materials Research Bulletin 47.2 (2012): 302-307. https://doi.org/10.1016/j.materresbull.2011.11.020

[72] Kaza, Jasmitha, Mallikarjuna Rao Pasumarthi, and P. S. Avadhani. "Superstrate and substrate thin film configuration of CdS/CZTS solar cell fabricated using SILAR method." Optics & Laser Technology 131 (2020): 106413. https://doi.org/10.1016/j.optlastec.2020.106413

[73] Krishnan, Ambily, et al. "Towards phase pure CZTS thin films by SILAR method with augmented Zn adsorption for photovoltaic applications." Materials for Renewable and Sustainable Energy 8 (2019): 1-8. https://doi.org/10.1007/s40243-019-0152-1

[74] Patil, B. M., et al. "Photo-electrochemical performance of Cu2ZnSnS4 thin films prepared via successive ionic layer adsorption and reaction method." Chemical Physics Letters 809 (2022): 140131. https://doi.org/10.1016/j.cplett.2022.140131

[75] Suryawanshi, M. P., et al. "A promising modified SILAR sequence for the synthesis of photoelectrochemically active Cu2ZnSnS4 (CZTS) thin films." Israel Journal of Chemistry 55.10 (2015): 1098-1102. https://doi.org/10.1002/ijch.201400203

[76] Suryawanshi, M. P., et al. "Improved photoelectrochemical performance of Cu2ZnSnS4 (CZTS) thin films prepared using modified successive ionic layer adsorption and reaction (SILAR) sequence." Electrochimica Acta 150 (2014): 136-145. https://doi.org/10.1016/j.electacta.2014.10.124

[77] Suryawanshi, M. P., et al. "CZTS based thin film solar cells: a status review." Materials Technology 28.1-2 (2013): 98-109. https://doi.org/10.1179/1753555712Y.0000000038

[78] Berruet, Mariana, et al. "Highly-efficient superstrate Cu2ZnSnS4 solar cell fabricated low-cost methods." physica status solidi (RRL)-Rapid Research Letters 11.8 (2017): 1700144. https://doi.org/10.1002/pssr.201700144

Materials Research Forum LLC
https://doi.org/10.21741/9781644903032-4

[79] Kumar, Vipin, and Vandana Grace Masih. "Fabrication and characterization of screen-Printed Cu 2 ZnSnS 4 films for photovoltaic applications." Journal of Electronic Materials 48 (2019): 2195-2199. https://doi.org/10.1007/s11664-019-07053-5

[80] Zhang, Ying, et al. "Ink formulation, scalable applications and challenging perspectives of screen printing for emerging printed microelectronics." Journal of Energy Chemistry 63 (2021): 498-513. https://doi.org/10.1016/j.jechem.2021.08.011

[81] Zhou, Zhihua, et al. "Fabrication of Cu2ZnSnS4 screen printed layers for solar cells." Solar Energy Materials and Solar Cells 94.12 (2010): 2042-2045. https://doi.org/10.1016/j.solmat.2010.06.010

[82] Wang, Yu, et al. "Influence of sintering temperature on screen printed Cu2ZnSnS4 (CZTS) films." Journal of alloys and compounds 539 (2012): 237-241. https://doi.org/10.1016/j.jallcom.2012.06.069

[83] Chen, Qinmiao, et al. "Cu2ZnSnS4 solar cell prepared entirely by non-vacuum processes." Thin Solid Films 520.19 (2012): 6256-6261. https://doi.org/10.1016/j.tsf.2012.05.074

[84] Chen, Qin-Miao, et al. "Doctor-bladed Cu2ZnSnS4 light absorption layer for low-cost solar cell application." Chinese Physics B 21.3 (2012): 038401. https://doi.org/10.1088/1674-1056/21/3/038401

[85] Peksu, Elif, and Hakan Karaagac. "Preparation of CZTS thin films for the fabrication of ZnO nanorods based superstrate solar cells." Journal of Alloys and Compounds 884 (2021): 161124. https://doi.org/10.1016/j.jallcom.2021.161124

[86] Flynn, Brendan, et al. "Microwave assisted synthesis of Cu2ZnSnS4 colloidal nanoparticle inks." physica status solidi (a) 209.11 (2012): 2186-2194. https://doi.org/10.1002/pssa.201127734

[87] Varadharajaperumal, S., et al. "Morphology controlled n-Type TiO2 and stoichiometry adjusted p- type Cu2ZnSnS4 thin films for photovoltaic applications." Crystal Growth & Design 17.10 (2017): 5154-5162. https://doi.org/10.1021/acs.cgd.7b00632

[88] Najafi, Vahid, and Salimeh Kimiagar. "Cd-free Cu2ZnSnS4 thin film solar cell on a flexible substrate using nano-crystal ink." Thin Solid Films 657 (2018): 70-75. https://doi.org/10.1016/j.tsf.2018.05.013

[89] Peng, Chien-Yi, et al. "Flexible CZTS solar cells on flexible Corning® Willow® Glass substrates." 2014 IEEE 40th Photovoltaic Specialist Conference (PVSC). IEEE, 2014.

[90] López-Marino, Simón, et al. "Earth-abundant absorber based solar cells onto low weight stainless steel substrate." Solar energy materials and solar cells 130 (2014): 347-353. https://doi.org/10.1016/j.solmat.2014.07.030

[91] Walsh, Aron, et al. "Kesterite thin-film solar cells: Advances in materials modelling of Cu2ZnSnS4." Advanced Energy Materials 2.4 (2012): 400-409. https://doi.org/10.1002/aenm.201100630

[92] Fan, Ping, et al. "Over 10% Efficient Cu2CdSnS4 Solar Cells Fabricated from Optimized Sulfurization." Advanced Functional Materials 32.45 (2022): 2207470. https://doi.org/10.1002/adfm.202207470

[93] Gokmen, Tayfun, et al. "Band tailing and efficiency limitation in kesterite solar cells." Applied Physics Letters 103.10 (2013): 103506. https://doi.org/10.1063/1.4820250

[94] Ma, Suyu, et al. "Origin of band-tail and deep-donor states in Cu2ZnSnS4 solar cells and their suppression through Sn-poor composition." The journal of physical chemistry letters 10.24 (2019): 7929-7936. https://doi.org/10.1021/acs.jpclett.9b03227

[95] Chatterjee, Soumyo, and Amlan J. Pal. "A solution approach to p-type Cu2FeSnS4 thin-films and pn-junction solar cells: role of electron selective materials on their performance." Solar Energy Materials and Solar Cells 160 (2017): 233-240. https://doi.org/10.1016/j.solmat.2016.10.037

[96] Guo, Huafei, et al. "The fabrication of Cu2BaSnS4 thin film solar cells utilizing a maskant layer." Solar Energy 181 (2019): 301-307. https://doi.org/10.1016/j.solener.2019.02.007

[97] El Radaf, I. M., et al. "Junction parameters and electrical characterization of the Al/n-Si/cu 2 CoSnS 4/au Heterojunction." Journal of Electronic Materials 48 (2019): 6480-6486. https://doi.org/10.1007/s11664-019-07445-7

[98] Elsaeedy, H. I. "Growth, structure, optical and optoelectrical characterizations of the Cu 2 NiSnS 4 thin films synthesized by spray pyrolysis technique." Journal of Materials Science: Materials in Electronics 30 (2019): 12545-12554. https://doi.org/10.1007/s10854-019-01615-3

[99] Das, Sonali, and Pitamber Mahanandia. "Improved PCE of solution processed kesterite Ag2ZnSnS4 quantum dot photovoltaic cell." Materials Chemistry and Physics 281 (2022): 125878. https://doi.org/10.1016/j.matchemphys.2022.125878

[100] Crovetto, Andrea, et al. "Wide band gap Cu2SrSnS4 solar cells from oxide precursors." ACS Applied Energy Materials 2.10 (2019): 7340-7344. https://doi.org/10.1021/acsaem.9b01322

[101] Kumar, Mukesh, et al. "Strategic review of secondary phases, defects and defect-complexes in kesterite CZTS-Se solar cells." Energy & Environmental Science 8.11 (2015): 3134-3159. https://doi.org/10.1039/C5EE02153G

[102] Olekseyuk, I. D., I. V. Dudchak, and L. V. Piskach. "Phase equilibria in the Cu2S-ZnS-SnS2 system." Journal of alloys and compounds 368.1-2 (2004): 135-143. https://doi.org/10.1016/j.jallcom.2003.08.084

[103] Nagoya, Akihiro, et al. "Defect formation and phase stability of Cu 2 ZnSnS 4 photovoltaic material." Physical Review B 81.11 (2010): 113202. https://doi.org/10.1103/PhysRevB.81.113202

[104] Murugan, Anbazhagan, et al. "Effect of Zn on nanoscale quaternary Cu2ZnSnS4 thin film electrodes for high performance supercapacitors." Journal of Energy Storage 44 (2021): 103423. https://doi.org/10.1016/j.est.2021.103423

[105] Islam, Md Fakhrul, Nadhrah Md Yatim, and Mohd Azman Hashim. "A review of CZTS thin film solar cell technology." Journal of Advanced Research in Fluid Mechanics and Thermal Sciences 81.1 (2021): 73-87. https://doi.org/10.37934/arfmts.81.1.7387

Chapter 5

Current Trends in Quantum Dots Solar Cells

Bavani Thirugnanam[1], Sivakami. A[2*], Madhavan Jagannathan[1]

[1]Solar Energy Lab, Department of Chemistry, Thiruvalluvar University, Vellore, 632 115, India

[2] Department of Physics, Sri Eshwar College of Engineering, Coimbatore-641202, Tamil Nadu, India

* sivakamitce@gmail.com

Abstract

Quantum dots (QDs) are zero- dimensional semiconductor structure exhibits tremendous applications in transistors, LEDs, photovoltaic cells, DNA imaging and photodetectors. The theoretical efficiency of QDs in photovoltaic cells is 40% but practically is lesser than the dye-sensitized solar cells. There are many recent developments focused on the use of different components like QD sensitizer, counter electrode, photoanode, electrolyte in quantum dot Sensitized Solar Cells (QDSSCs). The various synthesis methods are also contributing to enhance the efficiency of QDSSCs, open-circuit voltage, short-circuit current density and fill factor. The easily tunable band gap of QDSSCs is showing major improvement in the photovoltaic fields. The first practical efficiency of QDSSCs is recorded as 0.12% but now it is improved as 18% and many researchers are analyzing different factors to improve the efficiency of QDSSCs comparable with other solar cells.

Keywords

Quantum Dots, Sensitizer, Solar Cells, Photovoltaic, Efficiency, Fill Factor

Graphical Abstract

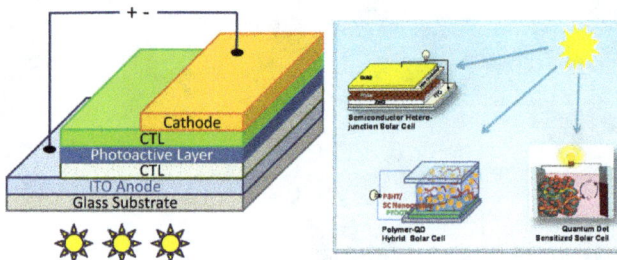

Contents

1.   Introduction

Currently, the production of sustainable and renewable energy with lesser effect on the environment, specifically the emission of CO_2 has considered as major challenge, to resolve this issue substantial assiduities were invested. The utilization of solar energy is considered as one of the most versatile and economical methods to produce renewable alternatives to fossil fuels. Various methods, including photovoltaic (PV) and photo electrochemical (PEC) methods are superior ways to produce renewable, transportable, and carbon free sources in the presence of solar energy [1-8]. Presently, a vital concentration of the research focuses on being cost effective and more stable with improved efficacy of the method, which generates highly effective solar cells. Presently, a broad variety of solar cell approaches have been researched and improved. Those are hybrid organic-inorganic solar cells, heterolayer solar cells and dye-sensitized solar cells. A single crystalline semiconductor wafer serves as the foundation for first-generation solar cells. In the cell assembly of second-generation photovoltaic solar cells, an organic thin film topology was applied [9-11]. Thin films, that usually have a coating over

Materials Research Forum LLC
https://doi.org/10.21741/9781644903032-5

optically conducting substrates and back electrodes, are the foundation of second-generation PV. The majorities of second-generation solar panels is built on CdTe and presently possess a 15% commercial rate. Whereas PV devices made from amorphous and nanocrystalline silicon as well as $CuInSe_2$, $CuInS_2$, $CuInGaSe_2$ have also achieved the level of marketing and involved to the PV market. The above thin film solar cells are cheaper to fabrication, than traditional ones, however their rate of conversion remains lower compared to that of first-generation single junction crystalline photovoltaic cells, which can reach up to 27% and is less than 14%. The maximum efficiencyof singlejunction cells ought to be able to show 33% [12-15]. Mono junction solar cells, that can surpass the Shockley-Queisser barrier of 31-41% energy productivity, are referred to as the third generation of PV cell technology. First-generation C-Si solar cells and thin-film solar cells possess various challenges that must be overcome thephotovoltaic technology to be developed in order to meet the demands of the golden triangle and achieve improved performances. Using advanced PV technologies including multi-junction cells, impurity band cells, multiple carrier generation via ionization, optical up- and downconverters, etc., third-generation solar cells apply to convert the energy more efficiently than the Shockley-Queisser limit [16-19]. The following solar cell kinds are composed of the third generation of solar cell technology:

1. Dye-Sensitized Solar Cells [DSSCs]

2. CZTS solar cells made of CZTSe and CZTSSe

3. Quantum Dot Sensitized Solar Cells [QDSSCs]

4. Perovskite photovoltaics

5. Organic photovoltaics

All the photons of incident sunlight must be absorbed by the solar cell in order to achieve the highest efficiency. Such excitonic solar cells perform in a way that's conceptually very similar as well. The geometric structures of them are varying considerably. Rather than typical free e^--h^+ pair found in traditional Si or other semiconductor driven P-N junction cells, these PV devices operate excitonic absorbent materials to generate excitons while absorbing incident solar irradiation. Because of their band offsets, most of the photo induced excitons disintegrates on the interface between e- h^+ and then migrate to the specific contacts [20-24].

Only the e- transport material (TM), that approves and transport of electron from the excitonic absorber, as well as the h^+ TM, which accepts and migrate the h^+ from the excitonic absorber which are employed to generate these selective interactions. Device efficiency could be a hampered by interface defects issues, energy level alignments, and carrier lifespan in e- TM and h+ TM. To obtain the improved photovoltaic response, rigorous material selection and preparation are crucial [25-28]. Nowadays, researchers have been interested in nanotechnology ever since their exceptional potential forthe application inoptoelectronicswas identified. The quantum confinement effect in low dimensional systems is governingall materialsphysical properties, when the size of nanocrystals is less than their corresponding Bohr exciton radius, these materials are known as quantum dots (QDs) [29-32]. A lot of research attention has been paid to QD recent years because of its exceptional opto-electronic features. In addition to PEC cells constructed from QD-sensitized wider-bandgap nanostructures, QD films dipped in electrolyte, solid-state cells driven QD/polymer blends, and QD layers bridged between e- and h+ conductors, and a variety of structures for QD-based solar cells have been reported in figure 1. It is simple to produce QD solar cells in a variety of methods and facilitating waveguide interaction and self-light monitoring. In order to achieve third-generation PVs and attain efficiency of conversion over the Shockley-Queisser limit, QD-based devices were also proposed. Due to the versatility in the absorption spectrum, single junction PV cells constructed using QDs absorber are viable candidates of being used as the base of third-generation multi-junction devices [32-41].

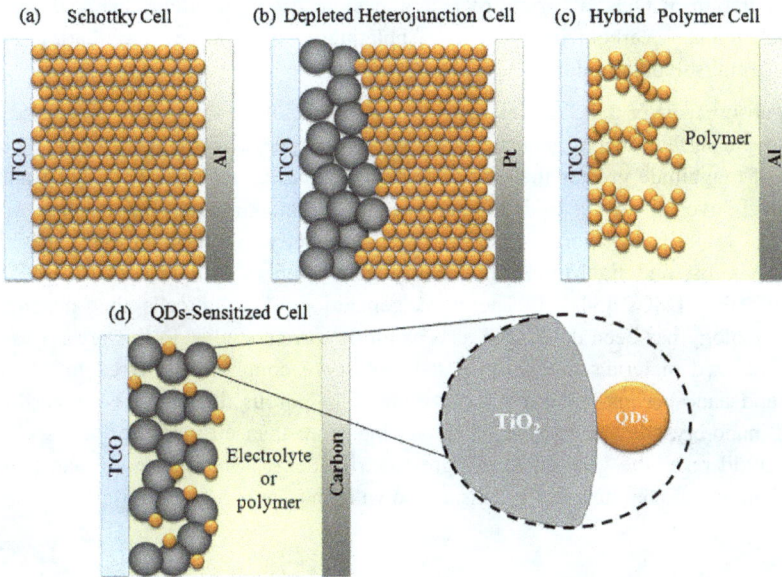

Figure 1. Different QDs based photovoltaic cells. Reproduced with permission from Ref. [41]

Herein, the researchers focus on current advances in wide-bandgap materials sensitized using QDs in QD solar cells (QDSCs), given the conceptual connection between QDSSCs and the relevance of the interface chemistry and physics is emphasized as an overview of the various QDSCs architectures and materials is provided, with a focus on PEC and solid-state QDSCs. A brief examination is given of third-generation PVs, alternative QD-based PV concepts, as well as advanced approaches. Figure 1 represents the different photovoltaic cells based on QDs.

1.1   Quantum-dot-sensitized solar cells

QDs, which are very small particles with a diameter of a few nanometers, are used in solar cells to absorb incident sunlight photons and produce the photovoltaic effect. The QDs have optical and electrical characteristics that are distinct from bulk material. The frequency of light released from QD can be finely controlled by tuningtheir size,andstructure [42-48]. Various size and type ofQDs generate light at varying frequencies, which frequency can possibly be precisely tuned to meet demands, QDs are additionally referred as artificial atoms because they contain aspects with genuine atoms or molecules, such as their singularity, bound structure, and distinct electronic states. For the preparation of QDs, a variety of techniques are used, including chemical ablating, electrochemical carbonizing, laser ablating, microwave irradiating, and hydrothermal/solvothermal treatment [49-53].

A thin absorber layer that has low absorbance, is capable of being used to sensitize the microscopic surface area of nanostructured wide-band gap semiconductor films, which is orders of magnitude greater than their overall geometrical regions. The light transmits that travels over up to hundreds of QD single layers making way for the poor optical density of a QD monolayer. The same principle underlying solar cell dye-sensitization and ETA cells, and light-to-electric energy generation rates in excess of 10% were attained using DSCs [54-57]. The third generation of solar cells that depends on nanotechnology has been developed as a result of current studies in this area. The most widely utilized materials are nano-crystals, or more commonly referred to "Quantum dots," and nano-porous materials like porous Si or porous titania, TiO_2. According to reports, nano-crystals are capable of converting more than 60% of the solar spectrum, which could result the generation of light-electric energy, which is more than twice as much than that of commercial PV cells as shown in the figure 5.2. [58-65].

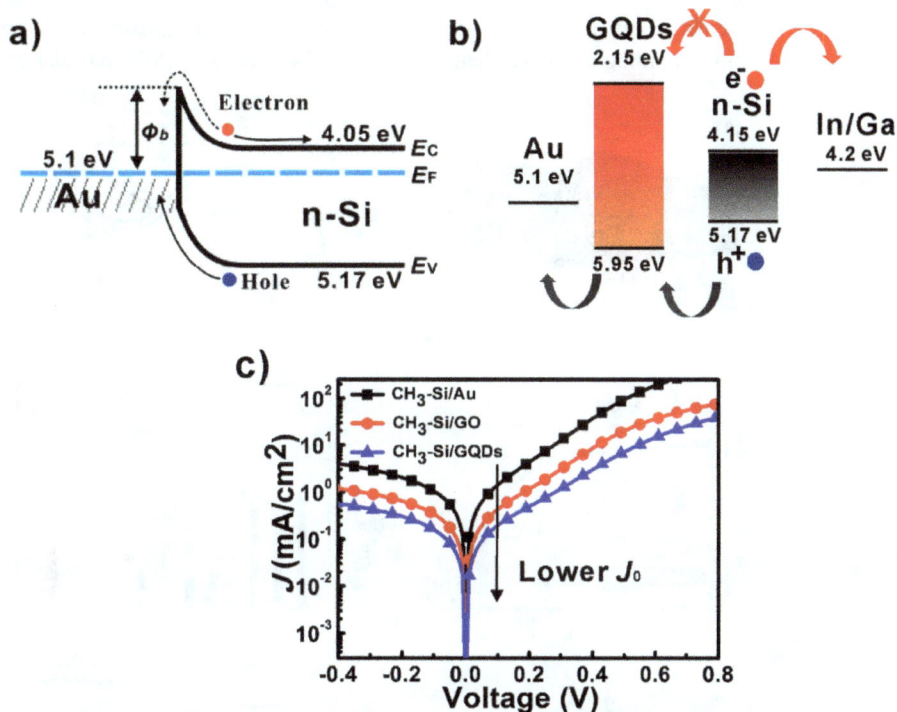

Figure 2. a) Energy band diagrams b) heterojunction device and c) V-I curve ofCH3−Si/Au, CH3−Si/GO, and CH3−Si/GQDs solar cells.Reproduced with permission from Ref. [65]

2. Recent developments in QDSSCs

2.1 Perovskite quantum dots

Organic-inorganic hybrid perovskites QDSC in PV applications have shown extraordinary success over the past decades, Zhang et al. [66] reported the highest efficiency 10.95% using spin coated $CsPbI_3$ perovskite film with 1-5 layers of ZnO as shown in figure 5.3. Among, 3 layerZnO coated $CsPBI_3$sample records highest efficiency and stability comparable with other layers of ZnO. From figure 5.4 concluded that the more layers of ZnOreduces the recombination and having higher ability of extracting charge carrier. Other researchers investigated on$CsPbI_3$and reported that the short-circuit current density and fill factor are greatly improved with post-treatment method optimization, resulting in an excellent efficiency of 14.10% for CsPbI$_3$ QD solar cells.

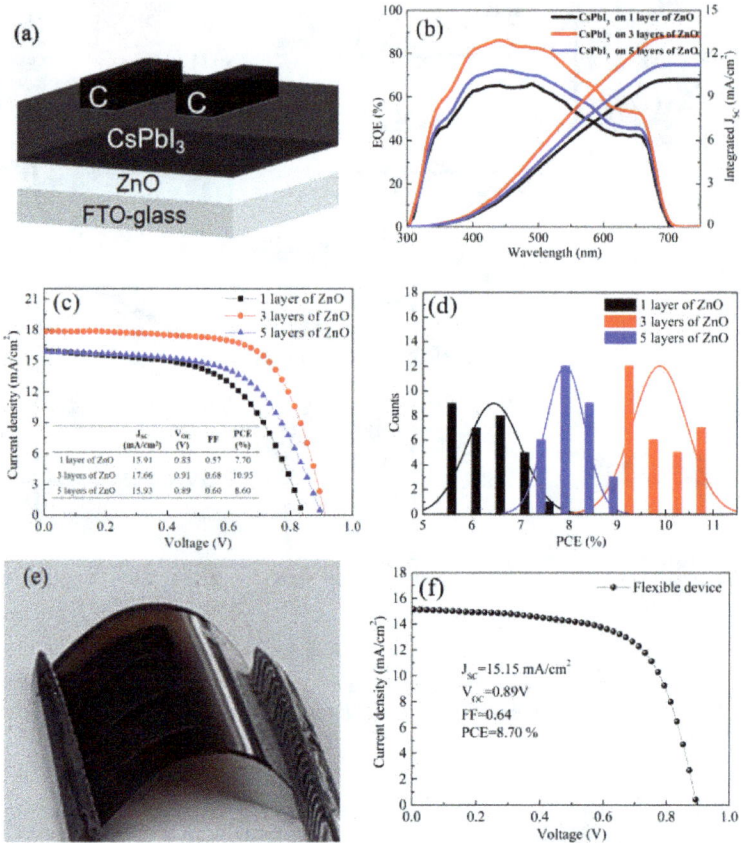

Figure 3. (a) The structure of the solar cell. (b) EQE spectra and the corresponding integrated JSC curves. (c) J-V curves. (d) Histograms of the CsPbI3 perovskite solar cells fabricated on different layer of ZnO. (e) The picture and (f) J-V curve of the flexible device based on 3 layers of ZnO. Reproduced with Permission from Ref.[66]

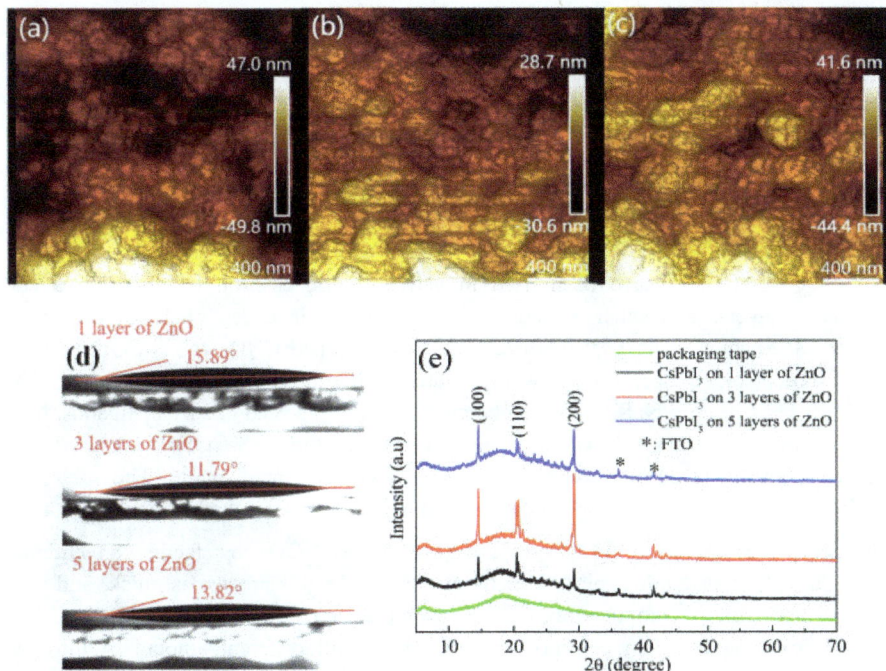

Figure. 4 AFM images of (a) 1 layer, (b) 3 layers, and (c) 5 layers of ZnO. (d) Contact angles of 1 layer, 3 layers and 5 layers of ZnO deposited on FTO substrates. (e) XRD patterns of the CsPbI3 films prepared on 1 layer, 3 layers and 5 layers of ZnO. Reproduced with permission from Ref. [66]

Xue et al. [67] adopting a systematically developed surface treatment technique, it was possible to successfully manage the surface ligand density and create theFAPbI₃ CQD-based photovoltaic device. The procedure involved using a series of "grade II" solvents (with moderate polarity that may efficiently reduce the ligand density while retaining the integrity of perovskite CQDs) with low polarity. With an extensive range of response up to the near-infrared region, our concept device displayed a promising energy conversion efficiency surpassing about 8%. It is important to note that compared to their bulk form, the FAPbI₃ CQD films and devices have demonstrated greater environmental and operating stability.

Zhang et al. created an internal P/N homojunction [68], effective CsPbI₃ perovskite QDSC having a layer of activity over 1 μm thick are reported. To generate various carrier-type QD arrays, 2,2'-(perfluoronaphthalene-2,6-diylidene) dimalononitrile

(F₆TCNNQ), an organic dopant, is incorporated to CsPbI₃ QD arrays. The P/N homojunction perovskite QDSC is subsequently joined together utilizing various carrier-type QDs resulting in an improved efficiency of power conversion of 15.29%. Furthermore, this P/N homojunction method achieves remarkable thickness sensitivity of QD solar cells, revealing enormous possibilities for future created by the generation of QDSC with an exceptional performance of 12.28% for a 1.2μ m-thick QD active-layer.

2.2   Carbon based quantum dots

Carbon QDs (CQDs) have been used in active layers, electron transporting layers, interfacial layers, and other structures because of their exceptional opto-electrical properties. Shahina et al. [69] reported S and N co-doped CQDs (N, S-CQDs)from a single precursor, thioacetamide, were created using a hydrothermal process. The synthesis is facile, and N, S-CQDs can be made without the need of surface ligands, high temperatures, or other complicated chemical processes, the N, S-CQDs were used as a "green" photoactive surface in a TiO₂ film as well as a photoanode. The solar cell produced a PCE of 1.36%, which is the greatest of all the QDSCs made of carbon.

Similarly, Darragh et al. [70] synthesized N-doped CQDs by facile conventional atmospheric pressure micro plasma method (figure.5.5), this process is quick, and eco-friendly and without complication, high temperature, and surface ligands. The N-CQDs exhibits exciting quantum confined optical assets with different amount of nitrogen combination. Significantly, the energy band structure of the N-CQDs is defended and they are incorporated into a photovoltaic device as the photoactive layer attaining a remarkable open-circuit voltage of 1.8 V and an efficiency of power conversion is 0.8%.

Jie et al. [71] designed zinc titanium mixed metal oxides derived from the layered double hydroxides precursor are employed as the passivation layers for CdS/CdSe co-sensitized solar cells. Herein, the addition of passivation layers significantly enhanced the efficiency of the corresponding CdS/CdSe QDs co-sensitized solar cells, achieving an excellent energy conversation about 4.91%, that's nearly 55% greater compared to that of devices without passivation. Suppression of charge carrier reconnection, increased e⁻ accretion effect, and improved surface characteristics have all been associated with the existence of ZnS/SiO₂ outer layers. Tian et al. [72] used a straightforward hydrothermal process and eco-friendly F127 soft templates; a honeycomb-shaped metallic 1T-MoS₂ for QDSSCs was successfully created. Electrochemical experiments reveal that 1T-MoS₂ has an improved electrocatalytic performance for Sn₂ reduction due to its interface charge transfer resistance (R_{ct}) is only 0.66 with 3% template. With Ti-mesh substrates MoS₂ QDSSCs' energy conversion performance achieved 6.03%. The exceptional efficiency can mainly attribute to 1T-MoS₂'s intrinsic conductivity, catalytic ability, and hydrophilicity as well as its distinctive structural feature in figure 5.6. The specific surface area of the honeycomb-shaped $1T\text{-}MoS_2$ is increased, and it also has extensive catalytic active sites and electrolyte transport ways to improve the material's stability.

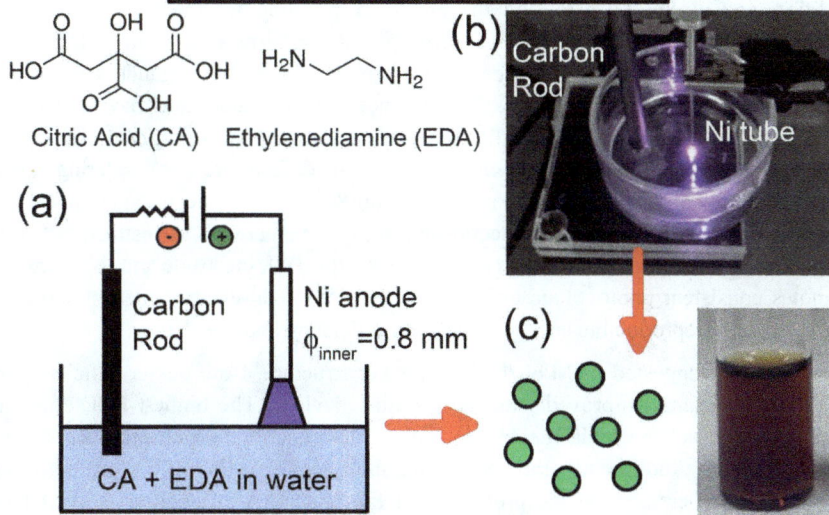

Figure 5. Microplasma synthesis of nitrogen-doped carbon dots (N-CQDs) from molecular precursors, (a) schematic of the setup used, (b) digital photograph of the microplasma–liquid interaction and (c) the N-CQDs Reproduced with permission from Ref. [70].

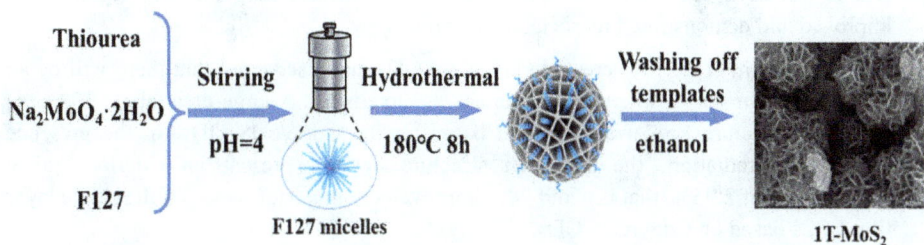

Figure 6. Schematic illustration of $1T\text{-}MoS_2$ preparation. [Reproduced with permission from Ref. [72].

Third Generation Photovoltaic Technology   Materials Research Forum LLC
Materials Research Foundations 163 (2024)118-144   https://doi.org/10.21741/9781644903032-5

2.3   Semiconductor quantum dots

SolarcellsconstructedusingZn:CuInSe$_2$QDsexhibitgoodefficiencywith   the   substantial
number of instinctive defects commonly of ternary I-III-VI2 semiconductors, Du et al
[73] conduct photovoltaic and spectroscopic analyses of as-prepared and surface-
modified Zn:CuInSe$_2$ QDs to find the factors responsible for the exceptional defect
tolerance of these types of devices. They adjusted the rates of both defect-related
relaxation and QD-to-TiO$_2$ electrode transfer of e- via surface ligands with various
lengths and binding affinities to the TiO$_2$ surface. Surface alterations have remarkably
little impact on photovoltaic efficiency while having a significant impact on
photoluminescence dynamics, showing that intragap defects are not hindering instead
assist the photoconversion process using Zn:CuInSe$_2$ QDs. These intragap states, that
have been considered shallow surface-located traps for e$^-$ as well as instinctive Cu$^+$ h$^+$-
trapping defects, control QD conversations between the TiO$_2$ electrode and the electrolyte
promotes consistent photovoltaic efficiency about 85% light-electric energy conversion
rate and greatly reproducible energy conversion efficiency about 9–10%.

Li et al. [74] Suggested a CsPbI$_3$/FAPbI$_3$ bilayer structure. Both device efficiency and
stability significantly improved with the resulting devices. The highest PCE efficiency
about 15.6% is achieved via the creation of a heterojunction, which provides superior
charge transport. Additionally, the ambient durability of the α-CsPbI$_3$/FAPbI$_3$ perovskite
QDSC greatly increased by the protection of the upper FAPbI$_3$ perovskite QD layer.
Also, Khan et al. [75] investigated the use of para-mercaptopyridine ligand post-
treatment on perovskite QDs and treatment of utilizing equivalent ortho-
mercaptopyridine and pyridine ligands process as shown in the figure 5.7 As a
consequence, the efficiency of CsPbI$_3$ perovskite QDSC was 14.25%. Additionally, after
around 70 days of storage at room temperature, the device's stability significantly
improved and demonstrated respectable performance.

Likewise, Gujju et al. [76] created CdSe/CdS QDs, anddiscovered that there will be an
enhancement in shape, structure and optical characteristics. A large-area (about 100 cm^2)
sandwich structure LSC was designed based on the CdSe/CdS QDs. In the presence
of sunlight irradiation, the sandwich structure devices present an exterior optical
performance of 2.95%, that is about 78% improvement in performance on the single layer
film LSCs based on CdSe/CdS QDs.

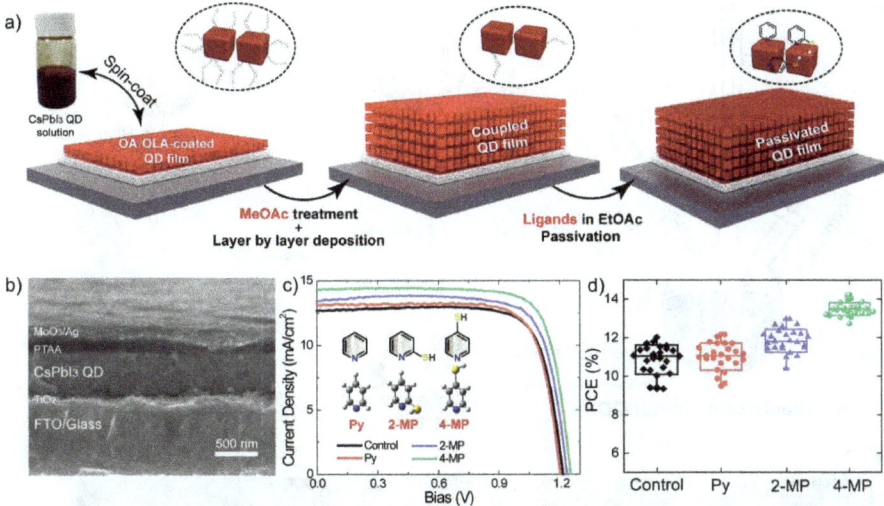

Figure 7. (a) Schematic diagram of the fabrication of CsPbI3 PQD solar cells; (b)crosssectional SEM image of the CsPbI3 PQD solar cell; (c) current density-voltage characteristics of the optimized CsPbI3 PQD solar cells. Reproduced with permission from Ref. [75]

2.4   Colloidal quantum dots solar cells

Colloidal QDs expresses effective multiple exciton effects, have size-dependent absorption and emission wavelengths and energy levels, and are surprisingly adjustable compounds reveal excellent durability. Hao et al. [77] presented an efficient cation-exchange method using oleic acid (OA) ligands to controllably synthesize $Cs_{1x}FA_xPbI_3$ QDs over the entire composition range (x= 0-1), which is not possible with large-grain polycrystalline thin films. The cross-exchange of cations was promoted in an OA-rich environment, allowing efficient fabrication of $Cs_{1x}FA_xPbI_3$ QDs with lower defect concentrations. With less hysteresis, the potential $Cs_{0.5}FA_{0.5}PbI_3$ QDSC achieves maximum PCE about 16.6%. The synthesis of colloidal QDS is shown in figure 5.8.

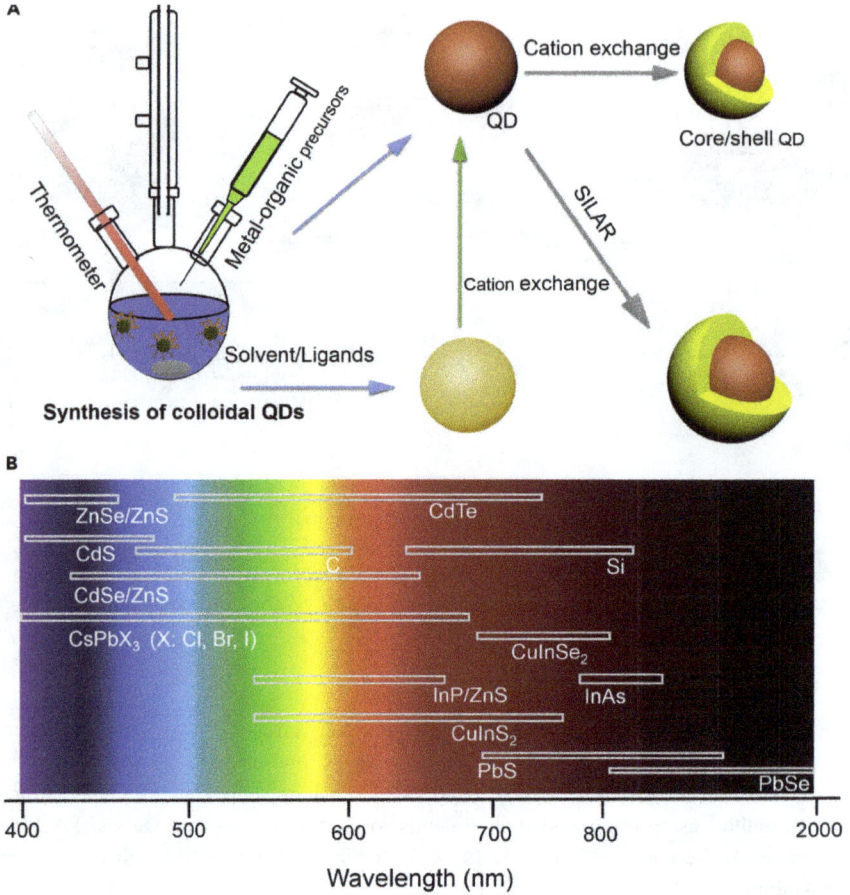

Figure 8. Schematic illustration of colloidal QDSSCs. Reproduced with permission from Ref. [78]

3. Developments in components of QDSSCs

3.1 QD sensitizer

QDs are absorber material for light harvesting in QDSSCs. Using different QD materials as QD sensitizer to improve the efficiency and reduce the recombination losses in QDSSCs. Those materials are CdTe, CdS, CdSe, PbS, InAS, CIS etc. Recent studies have

Materials Research Forum LLC
https://doi.org/10.21741/9781644903032-5

reported that employing two sensitizers are boosting the efficiency of QDSSCs by employing two sensitizers [79] and shown in the figure 5.9.

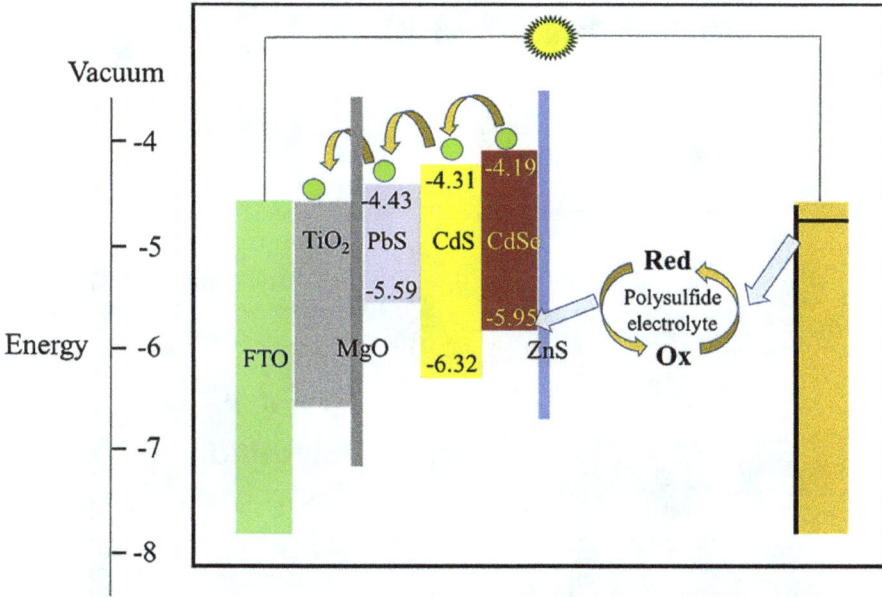

Figure 9. Schematic diagram with energy level cascade arrangement of PbS/ CdS/ CdSe QDSSCs with TiCl₄ and MgCl₂ treated TiO₂photoanode. Reproduced with permission from Ref. [79]

3.2   Photoanode developments in QDSSCs

The photoanodes play an essential role in electron transfer during recombination process. The nanostructured semiconductor metal oxides are referred as working electrode or photoanode for QDSSCs to improve the PCE. Those primary metal oxides are TiO_2, ZnO, SnO_2, $SrTiO_3$ etc.

Recently, Hami et al [80] fabricated and studied a TiO_2-MWCNTs nanocomposite filled mesoporous anatase TiO_2 photoanode for a PbS QDSSCs and PCE of 5.6. % and represented as figure.5.10.

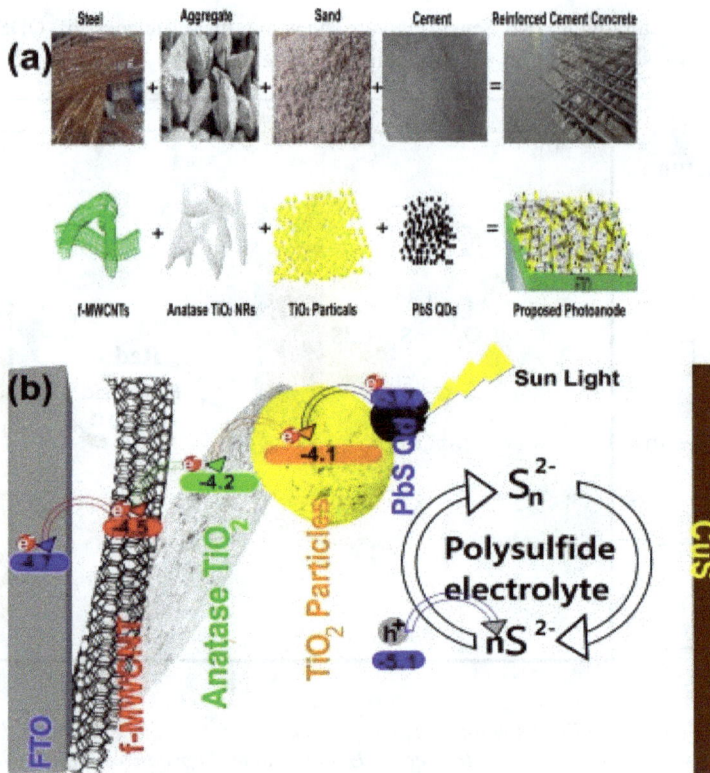

Figure 10. (a) Design analogy for conception of the photoanode, (b) Light excitation and electron transfer depiction. Reproduced with Permission from Ref.[80]

3.3 Electrolyte materials in QDSSCs

It is well known that the electrolyte plays a vital role to increase the efficiency and stability of all types of solar cells. There are many numbers of electrolytes used in QDSSCs but few are corrosive to QDSSCs to degrade the performance. So, three different electrolytes are used to improve the QDSSCs PCE namely, solid-state electrolytes, polymer gel electrolytes and quasi-state electrolytes [81] as shown in the figure 5.11.

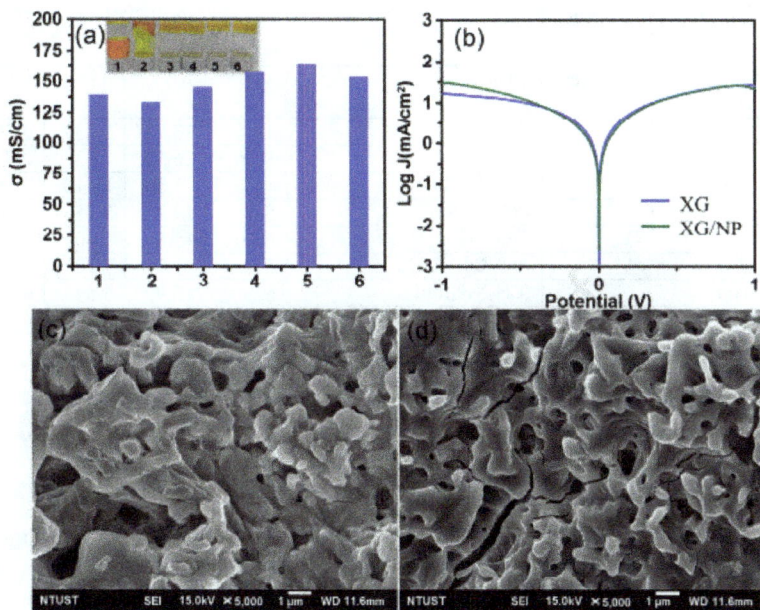

Figure 11. (a) Ionic conductivity of different concentrations of xanthan gum/NP electrolytes; The inset shows digital photograph of different polysulfide electrolyte: (1) Liquid electrolyte, (2) XG gel electrolyte, (3) 1 wt% TiO2 in XG gel electrolyte, (4) 2 wt% TiO2 in XG gel electrolyte, (5) 3 wt% TiO2 in XG gel electrolyte, and (6) 4 wt% TiO2 in XG gel electrolyte. (b) Tafel polarization curves of the symmetrical cells fabricated with two identical Cu2S electrodes and different polysulfide electrolytes at the scan rate of 50 mV/s. Typical SEM images of (c) XG gel and (d) XG/NP gel electrolytes. Reproduced with Permission from Ref. [81]

3.4 Developments in QDSSCs counter electrode

The working mechanism of counter electrode is to transfer the electron from circuit to electrolyte. The mobility of charge carriers is controlled by employing different counter electrode materials. It is well known that platinum (Pt) is a good catalyst used as counter electrodes for all types of solar cells. But Pt is more expensive so, there is a need to fabricate cost-effective counter electrode in the place of Pt. There will be many more decades of research in the fabrication of better counter electrodes. Recently, nitrogen doped carbon material was used as a CE material in QDSSCs and showed the PCE of 12.32% [82].

Table 1 shows that recent literatures different components and improvement of efficiency in QDSSCs.

S.No.	Photoanode (different metal oxides)	QDs	V_{OC}	J_{SC}	η (%)	F.F
1.	TiO_2	$CsPbI_3$	120	18.91	17.39	0.76 [83]
2.	TiO_2	Zn-Cu-In-S-Se	802	26.52	15.3	0.72 [84]
3.	SnO_2	$CsPbI_3$/PCBM	126	15.20	15.10	0.78 [85]
4.	N-TiO_2	CdS/CdSe	546	19.27	5.35	0.51 [86]
5.	TiO_2-fMWCNTS	PbS	650	15.80	5.60	0.55 [87]
6.	TiO_2	PbS/CdS/CdSe	520	11.77	7.70	0.53 [88]
7.	TiO_2	$(CH_3NH_3)PbI_3$	629	15.99	6.54	0.617 [89]
8.	TiO_2	Cd-, Pb-free Zn-Cu-In-Se	742	25.18	11.66	0.627 [90]

Summary and future prospects

QDSC possess an excellent possibility for becoming future generations of energy sources, due to its unique opto-electrical features, large surface area, excellent chemical stability, and excellent photostability.

Future research focuses on enhancing the efficiency of solar cells via raising light-harvesting performance of QDs by (1) the creation of completely "green" QDSCs. A need for an entirely novel kind of solar cell's practical usage requires that the entire system be ecologically friendly. (2) developing novel semiconductor QDs with a broad range of optical absorption, (3) light-harvesting efficiency (4) enhancing the electrical connection between QDs and e⁻ acceptors, (5) enhancing the mobility of electrons and device stability, discovering that the photovoltaic devices are stable over the long run. The stability of QDSCs today is still far below the level required for practical use. To increase the QDSCs' long-term stability, more work is needed. Making all-inorganic solid-state QDSCs or improving the device sealing method for liquid-junction QDSCs has the potential to produce devices with enough long-term stability (6) developing appropriate photoelectrode porosity to load more QDs and reduces the charge recombination and (7) reduced fabrication costs, QDSCs, however, are still in the early stages of formation and meet a lot of challenges. Furthermore, commercialization of large-scale solar cells based on node design is still not achieved. QDSSCs have the potential to substitute dye-sensitive solar cells, conserve materials for dual or thin-film solar cells, and contend with silicon solar cells in future generations. QDSCs have exciting advantages and rapid developments make them attractive candidates for the next generation of solar cells.

References

[1] Chen, J., Jia, D., Johansson, E.M., Hagfeldt, A. and Zhang, X., 2021. Emerging perovskite quantum dot solar cells: feasible approaches to boost performance. Energy Environ. Sci., 14(1), pp.224-261. https://doi.org/10.1039/D0EE02900A

[2] Yuan, J., Hazarika, A., Zhao, Q., Ling, X., Moot, T., Ma, W. and Luther, J.M., 2020. Metal halide perovskites in quantum dot solar cells: progress and prospects. Joule, 4(6), pp.1160-1185. https://doi.org/10.1016/j.joule.2020.04.006

[3] Albaladejo, Siguan, M., Baird, E.C., Becker-Koch, D., Li, Y., Rogach, A.L. and Vaynzof, Y., 2021. Stability of quantum dot solar cells: A matter of (life) time. Adv. Ener. Mater., 11(12), p.2003457. https://doi.org/10.1002/aenm.202003457

[4] Selopal, G.S., Zhao, H., Wang, Z.M. and Rosei, F., 2020. Core/shell quantum dots solar cells. Adv. Func. Mater., 30(13), p.1908762. https://doi.org/10.1002/adfm.201908762

[5] Lim, S., Han, S., Kim, D., Min, J., Choi, J. and Park, T., 2023. Key factors affecting the stability of CsPbI3 perovskite quantum dot solar cells: a comprehensive review. Adv. Mater., 35(4), p.2203430. https://doi.org/10.1002/adma.202203430

[6] Dias, J.A., Santagneli, S.H., Ribeiro, S.J. and Messaddeq, Y., 2021. Perovskite Quantum Dot Solar Cells: An Overview of the Current Advances and Future Perspectives. Solar RRL, 5(8), p.2100205. https://doi.org/10.1002/solr.202100205

[7] Sahu, A., Garg, A. and Dixit, A., 2020. A review on quantum dot sensitized solar cells: Past, present and future towards carrier multiplication with a possibility for higher efficiency. Solar Energy, 203, pp.210-239. https://doi.org/10.1016/j.solener.2020.04.044

[8] Rasal, A.S., Yadav, S., Kashale, A.A., Altaee, A. and Chang, J.Y., 2022. Stability of quantum dot-sensitized solar cells: A review and prospects. Nano Energy, 94, p.106854. https://doi.org/10.1016/j.nanoen.2021.106854

[9] Mahalingam, S., Manap, A., Omar, A., Low, F.W., Afandi, N.F., Chia, C.H. and Abd Rahim, N., 2021. Functionalized graphene quantum dots for dye-sensitized solar cell: Key challenges, recent developments and future prospects. Renew. Sust. Energ. Rev., 144, p.110999. https://doi.org/10.1016/j.rser.2021.110999

[10] Mora-Seró, I., 2020. Current challenges in the development of quantum dot sensitized solar cells. Adv. Energ. Mater., 10(33), p.2001774. https://doi.org/10.1002/aenm.202001774

[11] Kim, T., Lim, S., Yun, S., Jeong, S., Park, T. and Choi, J., 2020. Design strategy of quantum dot thin-film solar cells. Small, 16(45), p.2002460. https://doi.org/10.1002/smll.202002460

[12] Shaikh, J.S., Shaikh, N.S., Mali, S.S., Patil, J.V., Beknalkar, S.A., Patil, A.P., Tarwal, N.L., Kanjanaboos, P., Hong, C.K. and Patil, P.S., 2019. Quantum Dot Based Solar Cells: Role of Nanoarchitectures, Perovskite Quantum Dots, and Charge-Transporting Layers. ChemSusChem, 12(21), pp.4724-4753. https://doi.org/10.1002/cssc.201901505

[13] Blachowicz, T. and Ehrmann, A., 2020. Recent developments of solar cells from PbS colloidal quantum dots. Appl. Sci., 10(5), p.1743. https://doi.org/10.3390/app10051743

[14] Chen, M., Wang, J., Yin, F., Du, Z., Belfiore, L.A. and Tang, J., 2021. Strategically integrating quantum dots into organic and perovskite solar cells. J. Mater. Chem. A, 9(8), pp.4505-4527. https://doi.org/10.1039/D0TA11336K

[15] Ding, S., Hao, M., Lin, T., Bai, Y. and Wang, L., 2022. Ligand engineering of perovskite quantum dots for efficient and stable solar cells. J Energ. Chem., 69, pp.626-648. https://doi.org/10.1016/j.jechem.2022.02.006

[16] Zhao, Q., Hazarika, A., Chen, X., Harvey, S.P., Larson, B.W., Teeter, G.R., Liu, J., Song, T., Xiao, C., Shaw, L. and Zhang, M., 2019. High efficiency perovskite quantum dot solar cells with charge separating heterostructure. Nature commun., 10(1), p.2842. https://doi.org/10.1038/s41467-019-10856-z

[17] Guo, X., Zhao, B., Xu, K., Yang, S., Liu, Z., Han, Y., Xu, J., Xu, D., Tan, Z. and Liu, S., 2021. p-Type Carbon Dots for Effective Surface Optimization for Near-Record-Efficiency CsPbI2Br Solar Cells. Small, 17(37), p.2102272.

[18] Li, F., Zhou, S., Yuan, J., Qin, C., Yang, Y., Shi, J., Ling, X., Li, Y. and Ma, W., 2019. Perovskite quantum dot solar cells with 15.6% efficiency and improved stability enabled by an α-CsPbI3/FAPbI3 bilayer structure. ACS Energ. Lett., 4(11), pp.2571-2578. https://doi.org/10.1021/acsenergylett.9b01920

[19] Sivakami A, Gayathri.V, 2014, Effect of Nanocrystalline TiO$_2$ Film Thickness on the Photovoltaic Performance of Dye-Sensitized Solar Cells, J. Adv. Phy., 3(2), pp.119-124. https://doi.org/10.1166/jap.2014.1114

[20] Tavakoli, M.M., Dastjerdi, H.T., Yadav, P., Prochowicz, D., Si, H. and Tavakoli, R., 2021. Ambient stable and efficient monolithic tandem perovskite/PbS quantum dots solar cells via surface passivation and light management strategies. Adv. Funct. Mater., 31(21), p.2010623. https://doi.org/10.1002/adfm.202010623

[21] Cheng, F., He, R., Nie, S., Zhang, C., Yin, J., Li, J., Zheng, N. and Wu, B., 2021. Perovskite quantum dots as multifunctional interlayers in perovskite solar cells with dopant-free organic hole transporting layers. J. American Chem. Society, 143(15), pp.5855-5866. https://doi.org/10.1021/jacs.1c00852

[22] Xue, J., Wang, R., Chen, L., Nuryyeva, S., Han, T.H., Huang, T., Tan, S., Zhu, J., Wang, M., Wang, Z.K. and Zhang, C., 2019. A small molecule "charge driver" enables perovskite quantum dot solar cells with efficiency approaching 13%. Adv. Mater., 31(37), p.1900111. https://doi.org/10.1002/adma.201900111

[23] Han, R., Zhao, Q., Su, J., Zhou, X., Ye, X., Liang, X., Li, J., Cai, H., Ni, J. and Zhang, J., 2021. Role of methyl acetate in highly reproducible efficient CsPbI3 perovskite quantum dot solar cells. J.Phys. Chem. C, 125(16), pp.8469-8478. https://doi.org/10.1021/acs.jpcc.0c09057

[24] Pezhooli, N., Rahimi, J., Hasti, F. and Maleki, A., 2022. Synthesis and evaluation of composite TiO$_2$@ ZnO quantum dots on hybrid nanostructure perovskite solar cell. Scientific Reports, 12(1), p.9885. https://doi.org/10.1038/s41598-022-13903-w

[25] Xu, Z., Jiang, Y., Li, Z., Chen, C., Kong, X., Chen, Y., Zhou, G., Liu, J.M., Kempa, K. and Gao, J., 2021. Rapid microwave-assisted synthesis of SnO2 quantum dots for efficient planar perovskite solar cells. ACS Appl. Energy Mater., 4(2), pp.1887-1893. https://doi.org/10.1021/acsaem.0c02992

[26] Liu, H., Chen, Z., Wang, H., Ye, F., Ma, J., Zheng, X., Gui, P., Xiong, L., Wen, J. and Fang, G., 2019. A facile room temperature solution synthesis of SnO2 quantum dots for perovskite solar cells. J Mater. Chem. A, 7(17), pp.10636-10643. https://doi.org/10.1039/C8TA12561A

[27] A Sivakami, R Sarankumar, P Sudhagar, 2022, Graphene–Metal Oxides Nanocomposite Heterojunction as an Efficient Photocatalyst for Energy and Environmental Applications, Heterojunction photocatalytic materials, 29, I[st] edition Jenny stanford publishers, https://doi.org/10.1201/9781003294054

[28] Izmir, M., Durmusoglu, E.G., Sharma, M., Shabani, F., Isik, F., Delikanli, S., Sharma, V.K. and Demir, H.V., 2023. Near-infrared emission from CdSe-based nanoplatelets induced by ytterbium doping. J. Phys. Chem. C, 127(8), pp.4210-4217. https://doi.org/10.1021/acs.jpcc.2c09075

[29] Chaudhary, B., Kshetri, Y.K., Kim, H.S., Lee, S.W. and Kim, T.H., 2021. Current status on synthesis, properties and applications of CsPbX3 (X= Cl, Br, I) perovskite quantum dots/nanocrystals. Nanotechnology, 32(50), p.502007. https://doi.org/10.1088/1361-6528/ac2537

[30] Zheng, S., Chen, J., Johansson, E.M. and Zhang, X., 2020. PbS colloidal quantum dot inks for infrared solar cells. Isci., 23(11). https://doi.org/10.1016/j.isci.2020.101753

[31] Gan, J., He, J., Hoye, R.L., Mavlonov, A., Raziq, F., MacManus-Driscoll, J.L., Wu, X., Li, S., Zu, X., Zhan, Y. and Zhang, X., 2019. α-CsPbI3 colloidal quantum dots: synthesis, photodynamics, and photovoltaic applications. ACS Energy Lett., 4(6), pp.1308-1320. https://doi.org/10.1021/acsenergylett.9b00634

[32] Bi, C., Sun, X., Huang, X., Wang, S., Yuan, J., Wang, J.X., Pullerits, T. and Tian, J., 2020. Stable CsPb1-x Zn x I3 Colloidal Quantum Dots with Ultralow Density of Trap States for High-Performance Solar Cells. Chem. Mater., 32(14), pp.6105-6113. https://doi.org/10.1021/acs.chemmater.0c01750

[33] Yang, F., Xu, Y., Gu, M., Zhou, S., Wang, Y., Lu, K., Liu, Z., Ling, X., Zhu, Z., Chen, J. and Wu, Z., 2018. Synthesis of cesium-dopedZnO nanoparticles as an electron extraction layer for efficient PbS colloidal quantum dot solar cells. J. Mater Chem. A, 6(36), pp.17688-17697. https://doi.org/10.1039/C8TA05946B

[34] Ahmad, W., He, J., Liu, Z., Xu, K., Chen, Z., Yang, X., Li, D., Xia, Y., Zhang, J. and Chen, C., 2019. Lead selenide (PbSe) colloidal quantum dot solar cells with> 10% efficiency. Adv. Mater., 31(33), p.1900593. https://doi.org/10.1002/adma.201900593

[35] Chen, H., Luo, Q., Liu, T., Tai, M., Lin, J., Murugadoss, V., Lin, H., Wang, J., Guo, Z. and Wang, N., 2020. Boosting multiple interfaces by co-doped graphene quantum dots for high efficiency and durability perovskite solar cells. ACS Appl. Mater. Interfaces, 12(12), pp.13941-13949. https://doi.org/10.1021/acsami.9b23255

[36] Pang, S., Zhang, C., Zhang, H., Dong, H., Chen, D., Zhu, W., Xi, H., Chang, J., Lin, Z., Zhang, J. and Hao, Y., 2020. Boosting performance of perovskite solar cells with Graphene quantum dots decorated SnO2 electron transport layers. Appl. Surf. Sci.e, 507, p.145099. https://doi.org/10.1016/j.apsusc.2019.145099

[37] Wang, Z., Rong, X., Wang, L., Wang, W., Lin, H. and Li, X., 2020. Dual role of amino-functionalized graphene quantum dots in NiOx films for efficient inverted flexible perovskite solar cells. ACS Appl. Mater. Interfaces, 12(7), pp.8342-8350. https://doi.org/10.1021/acsami.9b22471

[38] Bian, H., Wang, Q., Yang, S., Yan, C., Wang, H., Liang, L., Jin, Z., Wang, G. and Liu, S.F., 2019. Nitrogen-doped graphene quantum dots for 80% photoluminescence quantum yield for inorganic γ-CsPbI3 perovskite solar cells with efficiency beyond 16%. J. Mater. Chem. A, 7(10), pp.5740-5747. https://doi.org/10.1039/C8TA12519H

[39] Zhou, Y., Yang, S., Yin, X., Han, J., Tai, M., Zhao, X., Chen, H., Gu, Y., Wang, N. and Lin, H., 2019. Enhancing electron transport via graphene quantum dot/SnO$_2$ composites for efficient and durable flexible perovskite photovoltaics. J. Mater. Chem. A, 7(4), pp.1878-1888. https://doi.org/10.1039/C8TA10168J

[40] Chen, X., Zhuang, Y., Shen, Q., Cao, X., Yang, W. and Yang, P., 2021. In situ synthesis of Ti$_3$C$_2$Tx MXene/CoS nanocomposite as high performance counter electrode materials for quantum dot-sensitized solar cells. Solar Energy, 226, pp.236-244. https://doi.org/10.1016/j.solener.2021.08.053

[41] Emin, S., Singh, S.P., Han, L., Satoh, N. and Islam, A., 2011. Colloidal quantum dot solar cells. Solar Energy, 85(6), pp.1264-1282. https://doi.org/10.1016/j.solener.2011.02.005

[42] Khan, F., Oh, M. and Kim, J.H., 2019. N-functionalized graphene quantum dots: Charge transporting layer for high-rate and durable Li4Ti5O12-based Li-ion battery. Chem. Eng. J, 369, pp.1024-1033. https://doi.org/10.1016/j.cej.2019.03.161

[43] Silambarasan, K., Harish, S., Hara, K., Archana, J. and Navaneethan, M., 2021. Ultrathin layered MoS2 and N-doped graphene quantum dots (N-GQDs) anchored reduced graphene oxide (rGO) nanocomposite-based counter electrode for dye-sensitized solar cells. Carbon, 181, pp.107-117. https://doi.org/10.1016/j.carbon.2021.01.162

[44] Sajjadi, S., Khataee, A., Soltani, R.D.C. and Hasanzadeh, A., 2019. N, S co-doped graphene quantum dot-decorated Fe3O4 nanostructures: Preparation, characterization and catalytic activity. J. Phys Chem. of Solids, 127, pp.140-150. https://doi.org/10.1016/j.jpcs.2018.12.014

[45] Mirtchev, P., Henderson, E.J., Soheilnia, N., Yip, C.M. and Ozin, G.A., 2012. Solution phase synthesis of carbon quantum dots as sensitizers for nanocrystalline TiO2 solar cells. J. Mater. Chem., 22(4), pp.1265-1269. https://doi.org/10.1039/C1JM14112K

[46] Hao, X.J., Cho, E.C., Flynn, C., Shen, Y.S., Park, S.C., Conibeer, G. and Green, M.A., 2009. Synthesis and characterization of boron-doped Si quantum dots for all-Si quantum dot tandem solar cells. Solar Energ Mater. Solar Cells, 93(2), pp.273-279. https://doi.org/10.1016/j.solmat.2008.10.017

[47] Ali, H.H. and Al-Bahrani, M.R., 2020. Synthesis of TiO$_2$/graphene quantum dots as photoanode to enhance power conversion efficiency for dye-sensitized solar cells. Int. J. Adv. Sci. Technol., 29(3), pp.11071-11081.

[48] Ganguly, A. and Nath, S.S., 2020. Mn-doped CdS quantum dots as sensitizers in solar cells. Mater. Sci. Eng. B, 255, p.114532. https://doi.org/10.1016/j.mseb.2020.114532

[49] Ali, H.H. and Al-Bahrani, M.R., 2020. Synthesis of TiO2/graphene quantum dots as photoanode to enhance power conversion efficiency for dye-sensitized solar cells. Int. J. Adv. Sci. Technol., 29(3), pp.11071-11081.

[50] Li, F., Zhou, S., Yuan, J., Qin, C., Yang, Y., Shi, J., Ling, X., Li, Y. and Ma, W., 2019. Perovskite quantum dot solar cells with 15.6% efficiency and improved stability enabled by an α-CsPbI3/FAPbI3 bilayer structure. ACS Energy Lett., 4(11), pp.2571-2578. https://doi.org/10.1021/acsenergylett.9b01920

[51] Hui, W., Yang, Y., Xu, Q., Gu, H., Feng, S., Su, Z., Zhang, M., Wang, J., Li, X., Fang, J. and Xia, F., 2020. Red-carbon-quantum-dot-doped SnO2 composite with enhanced electron mobility for efficient and stable perovskite solar cells. Adv. Mater., 32(4), p.1906374. https://doi.org/10.1002/adma.201906374

[52] Wang, X., Zhang, Y., Li, J., Liu, G., Gao, M., Ren, S., Liu, B., Zhang, L., Han, G., Yu, J. and Zhao, H., 2022. Platinum cluster/carbon quantum dots derived graphene heterostructured carbon nanofibers for efficient and durable solar-driven electrochemical hydrogen evolution. Small Methods, 6(4), p.2101470. https://doi.org/10.1002/smtd.202101470

[53] Maxim, A.A., Sadyk, S.N., Aidarkhanov, D., Surya, C., Ng, A., Hwang, Y.H., Atabaev, T.S. and Jumabekov, A.N., 2020. PMMA thin film with embedded carbon quantum dots for post-fabrication improvement of light harvesting in perovskite solar cells. Nanomaterials, 10(2), p.291. https://doi.org/10.3390/nano10020291

[54] Liu, J., Dong, Q., Wang, M., Ma, H., Pei, M., Bian, J. and Shi, Y., 2021. Efficient planar perovskite solar cells with carbon quantum dot-modified spiro-MeOTAD as a composite hole transport layer. ACS Appl. Mater. Interfaces, 13(47), pp.56265-56272. https://doi.org/10.1021/acsami.1c18344

[55] Niu, Y., Tian, C., Gao, J., Fan, F., Zhang, Y., Mi, Y., Ouyang, X., Li, L., Li, J., Chen, S. and Liu, Y., 2021. Nb2C MXenes modified SnO2 as high quality electron transfer layer for efficient and stability perovskite solar cells. Nano Energy, 89, p.106455.

[56] Huang, P., Xu, S., Zhang, M., Zhong, W., Xiao, Z. and Luo, Y., 2020. Carbon quantum dots improving photovoltaic performance of CdS quantum dot-sensitized solar cells. Opt. Mater., 110, p.110535. https://doi.org/10.1016/j.optmat.2020.110535

[57] Li, H., Shi, W., Huang, W., Yao, E.P., Han, J., Chen, Z., Liu, S., Shen, Y., Wang, M. and Yang, Y., 2017. Carbon quantum dots/TiOx electron transport layer boosts efficiency of planar heterojunction perovskite solar cells to 19%. Nano Lett., 17(4), pp.2328-2335. https://doi.org/10.1021/acs.nanolett.6b05177

[58] Zhou, Q., Tang, S., Yuan, G., Zhu, W., Huang, Y., Li, S. and Lin, M., 2022. Tailored graphene quantum dots to passivate defects and accelerate charge extraction for all-inorganic CsPbIBr2 perovskite solar cells. J. Alloys Compd., 895, p.162529. https://doi.org/10.1016/j.jallcom.2021.162529

[59] Maxim, A.A., Sadyk, S.N., Aidarkhanov, D., Surya, C., Ng, A., Hwang, Y.H., Atabaev, T.S. and Jumabekov, A.N., 2020. PMMA thin film with embedded carbon quantum dots for post-fabrication improvement of light harvesting in perovskite solar cells. Nanomaterials, 10(2), p.291. https://doi.org/10.3390/nano10020291

[60] Kurukavak, Ç.K., Yılmaz, T., Toprak, A., Büyükbekar, A., Kuş, M. and Ersöz, M., 2022. Improved performance with boron-doped carbon quantum dots in perovskite solar cells. J. Alloys Compd., 927, p.166851. https://doi.org/10.1016/j.jallcom.2022.166851

[61] Chen, G., Gong, Z., Bin, X. and Agbolaghi, S., 2023. Cutting-edge stability in perovskite solar cells through quantum dot-covered P3HT nanofibers. Polymer-Plastics Technol Mater., 62(2), pp.162-176. https://doi.org/10.1080/25740881.2022.2100791

[62] Das, S., Sa, K., Alam, I. and Mahanandia, P., 2019. Enhancement of photocurrent in Cu2 ZnSnS4 quantum dot-anchored multi-walled carbon nanotube for solar cell application. J. Mater. Sci., 54, pp.8542-8555. https://doi.org/10.1007/s10853-019-03467-y

[63] Chen, Y., Qiu, Q., Wang, D., Lin, Y., Zou, X. and Xie, T., 2019. CuSe/CuxS as a composite counter electrode based on PN heterojunction for quantum dot sensitized solar cells. J. Power Sources, 413, pp.68-76. https://doi.org/10.1016/j.jpowsour.2018.12.027

[64] Ling, X., Zhou, S., Yuan, J., Shi, J., Qian, Y., Larson, B.W., Zhao, Q., Qin, C., Li, F., Shi, G.and Stewart, C., 2019. 14.1% CsPbI3 perovskite quantum dot solar cells via cesium cation passivation. Adv. Energy Mater., 9(28), p.1900721.

[65] Gao, P., Ding, K., Wang, Y., Ruan, K., Diao, S., Zhang, Q., Sun, B. and Jie, J., 2014.

[66] Crystalline Si/graphene quantum dots heterojunction solar cells. J. Phys. Chem. C, 118(10), pp.5164-5171. https://doi.org/10.1021/jp412591k

[67] Junsen Zhang, Cheng Wang, Hao Fu, Li Gong, Haiping He, Zhishan Fang, Conghua Zhou, Jianlin Chen, Zisheng Chao, Jincheng Fan, 2021, Low-temperature preparation achieving 10.95%-efficiency of hole-free and carbon-based all-inorganic CsPbI3 perovskite solar cells, Journal of Alloys and Compounds, Volume 862,pp.58454. https://doi.org/10.1016/j.jallcom.2020.158454

[68] Jingjing Xue, Jin-Wook Lee, Zhenghong Dai, Rui Wang, Selbi Nuryyeva, Michael E. Liao, Sheng-Yung Chang, Lei Meng, Dong Meng, Pengyu Sun, Oliver Lin, Mark S. Goorsky, Yang Yang,2018, Surface Ligand Management for Stable FAPbI3 Perovskite Quantum Dot Solar Cells, Joule, 2 (9),pp. 1866-1878. https://doi.org/10.1016/j.joule.2018.07

[69] Xuliang Zhang, Hehe Huang, Xufeng Ling, Jianguo Sun, Xingyu Jiang, Yao Wang, Di Xue, Lizhen Huang, Lifeng Chi, Jianyu Yuan, Wanli Ma, 2021, Homojunction Perovskite Quantum Dot Solar Cells with over 1 μm-Thick Photoactive Layer, Adv. Materials. 34 (2), 2105977. https://doi.org/10.1002/adma.202105977

[70] Riaz, S. and Park, S.J., 2022. Thioacetamide-derived nitrogen and sulfur co-doped carbonquantum dots for "green" quantum dot solar cells. J. Indus. Eng. Chem., 105, pp.111-120. https://doi.org/10.1016/j.jiec.2021.09.009

[71] Carolan, D., Rocks, C., Padmanaban, D.B., Maguire, P., Svrcek, V. and Mariotti, D., 2017, Environmentally friendly nitrogen-doped carbon quantum dots for next generation solar cells. Sust. Energy Fuels, 1(7), pp.1611-1619. https://doi.org/10.1039/C7SE00158D

[72] Jie, J., Zhang, W., Bello, I., Lee, C.S. and Lee, S.T., 2010,One-dimensional II-VI nanostructures: synthesis, properties and optoelectronic applications. Nano Today, 5(4), pp.313-336. https://doi.org/10.1016/j.nantod.2010.06.009

[73] Tian, Z., Chen, Q. and Zhong, Q., 2020. Honeycomb spherical 1T-MoS2 as efficient counter electrodes for quantum dot sensitized solar cells. Chem. Eng. J., 396, p.125374.https://doi.org/10.1016/j.cej.2020.125374

[74] Du, J., Singh, R., Fedin, I., Fuhr, A.S. and Klimov, V.I., 2020. Spectroscopic insights into high defect tolerance of Zn: CuInSe2 quantum-dot-sensitized solar cells, Nat. Energy, 5(5), pp.409-417. https://doi.org/10.1038/s41560-020-0617-6

[75] Li, F., Zhou, S., Yuan, J., Qin, C., Yang, Y., Shi, J., Ling, X., Li, Y. and Ma, W., 2019, Perovskite quantum dot solar cells with 15.6% efficiency and improved stability enabled by an α-CsPbI3/FAPbI3 bilayer structure. ACS Energy Lett., 4(11), pp.2571-2578. https://doi.org/10.1021/acsenergylett.9b01920

Materials Research Forum LLC
https://doi.org/10.21741/9781644903032-5

[76] Khan, J., Zhang, X., Yuan, J., Wang, Y., Shi, G., Patterson, R., Shi, J., Ling, X., Hu, L., Wu, T. and Dai, S., 2020. Tuning the surface-passivating ligand anchoring position enables phase robustness in CsPbI3 perovskite quantum dot solar cells. ACS Energy Lett., 5(10), pp.3322-3329. https://doi.org/10.1021/acsenergylett.0c01849

[77] Liu, G., Mazzaro, R., Wang, Y., Zhao, H. and Vomiero, A., 2019. High efficiency sandwich structure luminescent solar concentrators based on colloidal quantum dots, Nano Energ., 60, pp.119-126. https://doi.org/10.1016/j.nanoen.2019.03.038

[78] Hao, M., Bai, Y., Zeiske, S., Ren, L., Liu, J., Yuan, Y., Zarrabi, N., Cheng, N., Ghasemi, M.,Chen, P. and Lyu, M., 2020. Ligand-assisted cation-exchange engineering for high-efficiency colloidal Cs1−xFAxPbI3 quantum dot solar cells with reduced phase segregation, Nat. Energ., 5(1), pp.79-88. https://doi.org/10.1038/s41560-019-0535-7

[79] Guiju Liu, Haiguang Zhao, Feiyu Diao, Zhibin Linga and Yiqian Wang, 2018, Stable tandem luminescent solar concentrators based on CdSe/CdS quantum dots and carbon dots, J. of Mat. Chem., 6 (37), pp.10059-10066. https://doi.org/10.1039/C8TC02532K

[80] Hao, M., Bai, Y., Zeiske, S., Ren, L., Liu, J., Yuan, Y., Zarrabi, N., Cheng, N., Ghasemi, M., Chen, P., Lyu, M., He, D., Yun, J., Du, Y., Wang, Y., Ding, S., Armin, A., Meredith, P., Liu, G., Cheng, H. & Wang, L. (2020). Ligand-assisted cation-exchange engineering for high-efficiency colloidal Cs1−xFAxPbI3 quantum dot solar cells with reduced phase segregation. Nature Energy, 5 (1), 79-88. http://dx.doi.org/10.1038/s41560-019-0535-7

[81] Zhao, H., Rosei, F., 2017, Colloidal Quantum Dots for Solar Technologies, Chem, 3(2), pp. 229-258. https://doi.org/10.1016/j.chempr.2017.07.007

[82] Manjceevan, A., Bandara, J., 2018, Systematic stacking of PbS/CdS/CdSe multi-layered quantum dots for the enhancement of solar cell efficiency by harvesting wide solar spectrum, Electrochimica Acta, 271, pp.567-575. https://doi.org/10.1016/j.electacta.2018.03.193

[83] H. Latif, S. Ashraf, M. Shahid Rafique, A. Imtiaz, A. Sattar, S. Zaheer, S. Ammara Shabbir, A. Usman, 2020, A novel, PbS quantum dot-Sensitized solar cell structure with TiO2-fMWCNTS nano-composite filled meso-porous anatase TiO2 photoanode, Sol. Energy, 204, pp. 617-623. https://doi.org/10.1016/j.solener.2020.03.114

[84] Y. Wang, J.H. Zeng, G. Zhao, S-alkylbenzothiophenium-based solid-state electrolyte for efficient quantum-dot sensitized solar cells, Sol. Energy, 194 (2019), pp. 286-293, https://doi.org/10.1016/j.solener.2019.10.047

[85] S. Jiao, J. Du, Z. Du, D. Long, W. Jiang, Z. Pan, Y. Li, X. Zhong, 2017, Nitrogen-doped mesoporous carbons as counter electrodes in quantum dot sensitized solar cells with a conversion efficiency exceeding 12, J. Phys. Chem. Lett., 8 (3), pp. 559-564, https://doi.org/10.1021/acs.jpclett.6b02864

[86] M. Ostadebrahim, H. Dehghani, 2020, Improving the photovoltaic performance of CdSe$_{0.2}$S$_{0.8}$ alloyed quantum dot sensitized solar cells using CdMnSe outer quantum dot, Sol. Energy, 199, pp. 901-910, 10.1016/j.solener.2019.10.036. https://doi.org/10.1016/j.solener.2019.10.036

[87]A. Arivarasan, S. Bharathi, S. Ezhil Arasi, S. Arunpandiyan, M.S. Revathy, R. Jayavel, 2020, Investigations of rare earth doped CdTe QDs as sensitizers for quantum dots sensitized solar cells, J. Lumin., 219 , Article 116881, https://doi.org/10.1016/j.jlumin.2019.116881

[88] P. Boon-on, D.J. Singh, J.-B. Shi, M.-W. Lee, 2020, Bandgap tunable ternary Cd$_x$Sb2-yS3$-\delta$ nanocrystals for solar cell applications, ACS Omega, 5 (1), pp. 113-121, https://doi.org/10.1021/acsomega.9b01762

[89] S.B. Patel, J.V. Gohel, 2019, Quasi solid-state quantum dot-sensitized solar cells with polysulfide gel polymer electrolyte for superior stability, J. Solid State Electrochem., 23 (9), pp. 2657-2666, https://doi.org/10.1007/s10008-019-04365-8

[90] S. Sujinnapram, S. Moungsrijun, 2015, Additive SnO$_2$-ZnO composite photoanode for improvement of power conversion efficiency in dye-sensitized solar cell, Procedia Manuf., 2, pp. 108-112, https://doi.org/10.1016/j.promfg.2015.07.019

[91] Z. Pan, I. Mora-Seró, Q. Shen, H. Zhang, Y. Li, K. Zhao, J. Wang, X. Zhong, J. Bisquert, 2014, High-efficiency 'green' quantum dot solar cells, J. Am. Chem. Soc., 136 (25), pp. 9203-9210, https://doi.org/10.1021/ja504310w

[92] M. Graetzel, 2007, Photovoltaic and photoelectrochemical conversion of solar energy, Philos. Trans. A. Math. Phys. Eng. Sci., 365, pp. 993-1005, https://doi.org/10.1098/rsta.2006.1963

[93] J.-Y. Chang, L.-F. Su, C.-H. Li, C.-C. Chang, J.-M. Lin, 2012, Efficient 'green' quantum dot-sensitized solar cells based on Cu$_2$S-CuInS$_2$-ZnSe architecture, Chem. Commun., 48 (40), pp. 4848-4850, 10.1039/C2CC31229H https://doi.org/10.1039/c2cc31229h

Third Generation Photovoltaic Technology
Materials Research Foundations 163 (2024) 145-167

Materials Research Forum LLC
https://doi.org/10.21741/9781644903032-6

Chapter 6

Perovskite Solar Cells: Current Strategy and Future Perspective

Govindhasamy Murugadoss[1, *], Manogaran Rajasekaramoorthy[1] and Alagarsamy Pandikumar[2]

[1]Centre for Nanoscience and Nanotechnology, Sathyabama Institute of Science and Technology Chennai-600119, Tamil Nadu, India

[2]Electro Organic and Materials Electrochemistry Division, CSIR-Central Electrochemical Research Institute, Karaikudi 630003, Tamil Nadu, India

murugadoss_g@yahoo.com

Abstract

Perovskite solar cells (PSCs) have demonstrated notable improvements in their power conversion efficiency (PCE), indicating both academic research and commercial application value. The PCE gap is closing when compared to silicon cells sold in stores. Large-scale production, cost, and stability, however, are still far behind. Every functional layer of perovskite solar cells has a range of relevant research for scaling up the preparation of high-efficiency and stable PSCs. The functional layers, such as the electron transport layer, perovskite layer, hole transport layer, and electrode, have been the subject of recent research, which is systematically summarised in this chapter. Significant advancements in device stability and efficiency over the last few decades can be attributed to massive research efforts in compositional, process, and interfacial engineering. We discuss the benefits and drawbacks of PSCs in comparison to the current silicon photovoltaic technology with regard to commercial applications. Moreover, we discuss the structural stability, optical properties, perovskite device structure and operation principle, High efficiency PSCs, Perovskite powder production for diverse application. PSCs provide low manufacturing costs and solution processability, but on the road to commercialization, their poor stability and element toxicity need to be addressed. It is yet unknown how to resolve the costly and unstable issues with electrode materials and Spiro-OMeTAD. There is also discussion of the primary issues and the path for their future growth. In addition, we offer our predictions for PSC commercialization in the solar industry. PSCs are expected to show greater promise in tandem configurations and low-cost modules.

Keywords

Perovskite Film, Band Gap Tuning, Cubic Structure, Stability, Fast Crystallization, Scale-Up

Contents

1. Introduction

Perovskite solar cells (PSCs) are among the most promising emerging photovoltaic technologies due to their remarkable power conversion efficiency. Perovskite has the general formula of ABX3, where A is an organic or metal cation (e.g., MA^+, FA^+, Cs^+), B is a metal cation (e.g., Pb^{2+}, Sn^{2+}), and X is a halogen (e.g., Cl^-, Br^-, I^-). When compared to silicon, perovskite offers better optoelectronic qualities and a lower crystallisation activation energy [1-4]. Due to its exceptional electrical and optical qualities, perovskite has attracted a lot of attention in recent years and has a lot of potential for application in next-generation solar cells. The primary goal of photovoltaic (PV) products is to generate energy and ensure that it is appropriate for large-scale use. However, a number of difficulties that are not present in smaller devices are encountered because of the PV unit's size [5]. Because there could be no element oxide scaffold, perovskite solar cells (PSCs) were initially produced from DSSC experiments. The presence of interface layers, one of which is thought to be for holes and one of which is thought to be for electrons, is a basic prerequisite. These interface layers may be charge selective. PTAA, PEDOT, PSS, CuSCN, and others are recognised interface layers that perform significantly better in the domain of PSCs [6].

On the other hand, metal oxides, such as tin oxide, effectively serve as the electron transport layer for electron transporters [7]. Following light absorption, the photogenerated charge is utilised in the electron and hole interface layers via the perovskite layer. The typical structure contains an absorbent layer made of a perovskite substance that is 300–500 nm deep [8–9]. Over a big area, perovskites' morphology is difficult to control. The light absorption of perovskite materials may be influenced by their shape. The different fabrication processes have been researched in light of the improvement in power conversion efficiency (PCE) [10]. There is an additional method to improve the optoelectronic properties by changing the shape of the perovskite layer. Using a gas-assisted fabrication technique, Cheng et al. created a perovskite film with a surface texture that has a textured CH3NH3PbI3 morphology created by seeding a thin mesoporous TiO2 layer. This textured morphology has a multitiered nanostructure that significantly improves the solar cell's ability to extract charges and capture light [11]. According to Wang et al., a whispering-gallery (WG) framework for light-imitating entrapment was constructed on a perovskite active layer for antireflection and light-harvesting using merely an imprinting process and a sturdy microstructure stamp for PSCs. The optical feedback and consistent absorption capabilities of the WG-dependent perovskite films enable light trapping. Additionally, this process can hasten electron-hole separation and prevent recombination. The J-V hysteresis of the advanced WG device is hardly detectable. The quality and thickness of the film during manufacture are two more important factors that must be addressed. Careful consideration should be given to the manufacturing processes.

As a result, it is important to list the various fabrication procedures and the associated film morphology [6]. It is well known that the fabrication process has a direct impact on the morphology of perovskite films. We outlined the most recent manufacturing techniques and their impacts on the morphology of perovskite films in this publication. Additionally, the entire performance of the device, including stability and efficiency, was examined. Finally, depending on the various fabrication processes, the possibility of the commercialization of perovskite solar applications is examined. In lab tests, the one-step and two-step sequence deposition techniques are primarily researched to increase efficiency, demonstrate the superiority of PSCs, and draw more public attention. The accomplishments could aid in the advancement of PSC commercialization as they pertain to the fundamental study of PSCs in the lab.

Over the past few years, considerable progress has been accomplished in lab experiments thanks to enormous efforts. The efficiency of large-scale perovskite photovoltaic devices, however, is still much lower than that of perovskite devices of modest size (0.1 cm^2). To spur additional research, a summary of recent developments in large-scale fabrication technologies is required. As a result, we go into great detail about perovskite solar cells from the following perspectives: A flat and compact morphology for high-quality perovskite films is a requirement for high-efficiency devices. Vacuum deposition, one-step solution deposition (antisolvent dripping, solvent-assisted annealing, and precursor engineering), two-step sequential deposition, and vapor-assisted deposition are only a few of the processing methods that have been created to date to create high-quality perovskite films. Among these techniques, low-cost, solution processes that are compatible with flexible electronics show promise. Additionally, scaling-up techniques (such as inkjet printing, roll-to-roll coating, and slot-die coating) have been created for the future industrial production of PSCs.

2.   tructural stability

Crystal-structure transition is an essential factor that has gotten less attention among the many variables that potentially affect the stability of perovskite materials. Depending on the size and interaction between the A cation and the corner-sharing BX_6 octahedra, materials with an ABX_3 composition can adopt various crystal forms. A trustworthy empirical index for determining which structure will emerge preferentially is the Goldschmidt tolerance factor (t) [12]. The following expression can be used to compute the Goldschmidt tolerance factor from the ionic radius of the atoms: where r_A is the radius of the A cation, r_B is the radius of the B cation, and r_X is the radius of the anion

$$t = \frac{r_A + r_X}{\sqrt{2}\,(r_B + r_X)}$$

Materials that have a tolerance factor of 0.9 to 1.0 typically have a perfect cubic structure. A perovskite structure that is deformed and has slanted octahedra is produced by a tolerance factor of 0.71–0.9. When the tolerance factor is higher (>1) or lower (0.71), non-perovskite structures emerge [12]. The tendency is still true for inorganic–organic hybrid halide perovskite materials, even if the rule was created for oxide perovskite [13–14]. Figure 1 shows a schematic illustration of the ABX_3 perovskite's straightforward cubic structure. When t < 0.8, 0.8 < t < 1, and t > 1, the inorganic–organic hybrid halide perovskite materials often form orthorhombic, cubic, and hexagonal structures, respectively. In other words, halide perovskites exhibit a tendency towards lower symmetry involving tetragonal, orthorhombic, trigonal, and even hexagonal when the tolerance factor is less than 0.97. This tendency results from cooperative octahedron distortion. Depending on the temperature and fabrication techniques, a perovskite material with a specific chemical composition is typically found to have more than one structure. In this study, the phase with cubic structure is referred to as the -phase and the phase with non-perovskite structures as the δ-phase for ease of comparison. Additionally, the octahedral factor, which has the following expression [15], is a significant index.

$$\mu = \frac{R_B}{R_X},$$

with R_B and R_X as the ionic radii. It has been experimentally proven that an unstable perovskite can be synthesised with an octahedral factor lower than 0.442 even though the system has a reasonable tolerance factor [15]. The octahedral factor is typically in the range of $0.377 < \mu < 0.895$, but specifically for the formation of halide perovskites, the lowest limit of is actually 0.442. As a result, it is determined that the tolerance factor and octahedral factor are necessary but insufficient for the creation of halide perovskites. In other words, the aforementioned theoretical definitions can be used to infer crystallisation stability and potential structures. It is noteworthy to emphasize that constituent ions are treated as packed rigid spheres with radii of $r_{MA^+} = 1.8$ Å [16], $r_{FA^+} = 1.9–2.2$ Å [17], $r_{Pb^{2+}} = 1.19$ Å, $r_{I^-} = 2.2$ Å, $r_{Cl^-} = 1.81$ Å and $r_{Br^-} = 1.96$ Å [18]. It is obvious that the injection of large-size formamidinium cations or small-size Br/Cl ions causes the tolerance factor to approach 1, indicating that the stability is significantly improved by the aforementioned ion doping.

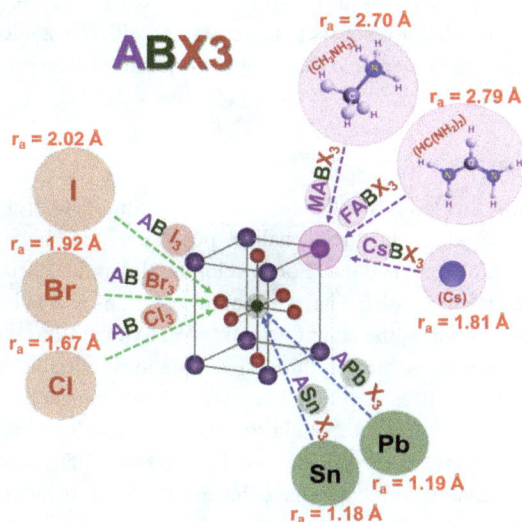

Figure 1. A simple cubic structure of the ABX₃ perovskite.

Third Generation Photovoltaic Technology   Materials Research Forum LLC
Materials Research Foundations 163 (2024) 145-167   https://doi.org/10.21741/9781644903032-6

The organic cation in inorganic–organic hybrid perovskite typically has a nonspherical geometry and spins continuously in the lattice [19–20], in contrast to inorganic perovskite. As a result, it might be challenging to estimate the precise size of the organic cation and compute the absolute tolerance factor for a given chemical. The structure transition in such materials can still be understood through qualitative investigation, though. For instance, compared to methylammonium lead iodide, $CH_3NH_3PbI_3$ (MAPbI$_3$), formamidinium lead iodide, $HC(NH_2)_2PbI_3$ (FAPbI$_3$), has a bigger A cation. The geometrical structure of the FA$^+$ cation makes it impossible to determine the exact tolerance factor; however, a larger cation would typically have a higher tolerance factor t. In FAPbI$_3$ materials that have undergone solution processing, two phases with different crystal structures have been obtained. One is the perovskite-structured, photoactive -phase (black phase), and the other is the nonphotoactive -phase (yellow phase), which we refer to as the δH-phase [21]. Given Goldschmidt's rule, it is plausible to conclude that the tolerance factor for hexagonal δH-FAPbI$_3$ is greater than 1. Inorganic perovskite solar cells recently showed respectable efficiency when using Cs$^+$ as the A cation [22–23]. Compared to their organic hybrid counterparts, inorganic halide perovskite materials frequently have a wider temperature range for thermal disintegration. Therefore, this increased design flexibility opens up new opportunities for producing stable PSCs. The tolerance factor is too low as a result of the small-size Cs$^+$ cation to support a cubic perovskite structure, though. At temperatures above 300 °C, the photoactive α-CsPbI$_3$ with a bandgap of 1.77 eV is often achieved [24]. However, the tolerance-factor analysis is in agreement with the stable phase of CsPbI$_3$ at room temperature, which is a nonphotoactive yellow phase with an orthorhombic structure (referred to as the "δ-phase").

3.   Optical properties

It was discovered in this study for the first time the complicated refractive indices (and dielectric properties) of organometal halide perovskites ($CH_3NH_3PbI_{(3-x)}Cl_x$) in the wavelength range of 300 to 1100 nm. For developing the optical structures of perovskite solar cells, these facts are crucial. The selection of electron- and hole-transporting layers (ETLs/HTLs), optimisation of the layer thickness of individual films, and carrier balancing of tandem cells were all achieved through simulations utilising the transfer matrix approach. The findings offer a broad overview of advantageous optical layouts for perovskite photovoltaics, which is thought to hasten the development of perovskite solar cells towards power conversion efficiencies (PCE) above 20%. The optical band gap, valence band, and conduction band have been calculated using measurements from reflectance and ultraviolet photoelectron spectroscopy (UPS), with the corresponding values being estimated at 1.5, 5.4, and 3.9 eV. Density functional theory (DFT) is then used to theoretically determine the electronic structure of perovskites without taking spin-orbit coupling into account. These findings support the high photovoltaic qualities of halide perovskites by agreeing with the experimental findings (1.5 eV). A schematic representation of the energy levels of electron and hole transporting materials (HTM) that are well aligned with the perovskite in Figure 2 results in high charge separation and transport.

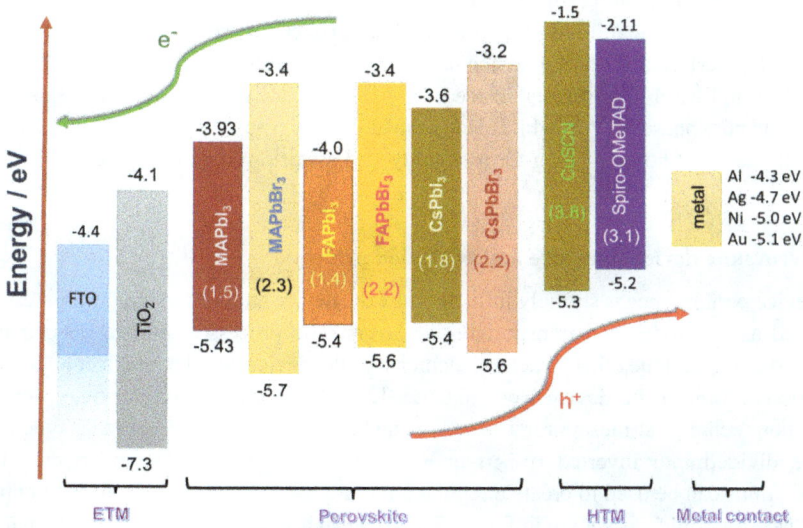

Figure 2 Energy level diagram for representative perovskite with electron transport layer, hole transport layer and contact layer.

Strong correlations exist between ion species and the optoelectronic characteristics of methyl ammonium halide perovskite (MHPs). First off, X-site halide mixing allows for continuous bandgap tuning over practically the whole visible spectrum [18,25]. When I is replaced with Br ions, the absorption edge shifts from 786 to 544 nm, causing the bandgap to rise from 1.58 to 2.28 eV. Wide-bandgap perovskites can be formed into tandem solar cells using an efficient technique called bandgap engineering by halide ions. The FA^+ ion is a significant replacement for the MA^+. The replacement enlarges the perovskite crystal and reduces the Pb-I bond distance due to the bigger ion size of FA^+, which leads to the lowered bandgap of 1.47 eV. The use of FA^+ as the A-site cation in MHPs further increases photocurrents.

Third Generation Photovoltaic Technology   Materials Research Forum LLC
Materials Research Foundations 163 (2024) 145-167   https://doi.org/10.21741/9781644903032-6

Unfortunately, according to the Goldschmidt rule, the following chemical management still affects the phase stability of FAPbI₃. The photoactive α-FAPbI3 is unstable at ambient temperatures and readily converts into the yellow δ phase, which significantly impairs the device's performance. To address this issue, new methods for stabilising the -phase in FA-based PSCs are required. By including MAPbBr₃, Seok and colleagues first stabilised -phase FAPbI₃ and attained a PCE of more than 18% PCE [4]. Saliba and his colleagues later mixed FA⁺, MA⁺, and Rb⁺ to further optimise the A site cation as a "cation cascade" and get rid of the remaining δ-phase [26]. By using this technique, FA-based PSCs are capable of providing a stabilised PCE of over 21% and a lengthy shelf life of over 1,000 h. However, the performance of the device may be constrained by the employment of MA⁺, Rb⁺, and Br⁻ in FAPbI₃ perovskites. Instead, the pseudo-halide anion (HCOO⁻) was used to stabilise the pure phase FAPbI₃ while also passivating the I⁻ vacancy defects. The resulting solar cells had 450 hours of long-term stability and a world-record efficiency of 25.6% [27].

4. Perovskite device structure and operation principle

PSC device performance is strongly influenced by the device architecture and the materials employed as electrodes and transport buffer layers. The processes used to deposit the photoactive film and the other structural elements of the device could also affect how well the various layers of the device work together [28]. The PSC uses two primary device preparation techniques: mesoporous structures and planar structures. The planar type can also be divided into inverted (p-i-n) and regular (n-i-p) forms. The n-i-p or p-i-n configurations can be used to create a regular structure based on the mesoporous titanium oxide (TiO₂) scaffold. As shown in Figure 3 [29], it can also be produced in simple planar regular or inverted designs.

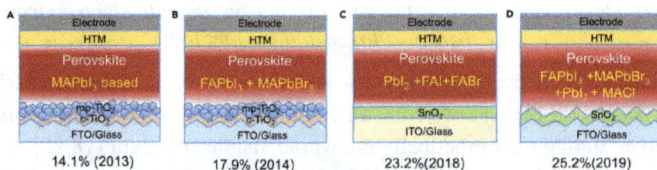

Figure 3. Changes in the structure and composition of PSCs according to the conversion efficiency record [30].

n-i-p Structure

The DSSC method, in which the perovskite layer absorber self-assembles within the porous region of TiO₂ layer, served as the basis for the first hybrid perovskite absorbers used in the construction of solar cells. In order to protect the perovskite film and transmit electrons from the conduction band (CB) of the perovskite to the compact TiO_2 layer, a mesoporous layer of metal oxide is also formed on the compact TiO_2 layer [31]. Therefore, the most extensively used device structure for creating extremely effective perovskite solar cells is the mesoscopic n-i-p design. By depositing a thin layer of TiO_2 (10–30 nm) as a hole-blocking layer on a conductive transparent oxide like fluorine-doped tin oxide (FTO) or indium tin oxide (ITO), the mesoporous layer is created. In order to allow the infiltration of perovskite precursor from the solution into the TiO_2 medium, a thick porous layer of TiO2 nanoparticles is present. In contrast, a perovskite film is formed up to a thickness of 300 nm on top of compact electron transport material (ETM) that is (50-70) nm thick. This is followed by the deposition of hole transport material (HTM), which is (50-100) nm thick. In planar designs, however, there is no mesoporous layer. The device is finished by deposition of the metal anode. The top contact electrode can be completely hidden in this configuration when light passes through the glass substrate and then into the hole carrying material, while reflection from the top contact can serve to increase PCE [32].

p-i-n Structure (inverted)

Inverted architecture, or p-i-n, is prepared by sequential deposition of the various layers of materials. Initially, a 50 to 80-nm-thick HTM, p-type conductive polymer poly(3,4-ethylenedioxy thiophene) polystyrene sulfonate (PEDOT:PSS) is deposited on ITO coated glass substrates, followed by up to 300-nm intrinsic perovskite thin-film. Finally, the device is completed with the deposition of (10–60) nm organic hole-blocking layer such as-phenyl-C61-butyric acid methyl ester (PCBM) and a metal cathode (Al or Au) [33]. In order to increase carrier diffusion length in the manufacturing of mixed halide organolead trihalide perovskites, Dong et al. reported mixed-halide perovskite via a multi-cycle coating. A 18.9% was the highest PCE of a planar p-i-n solar cell ever recorded [34]. The p-i-n device structure also provides more alternatives for ETMs, including both organic and inorganic compounds. For the hole- and electron-selective connections, respectively, NiO and ZnO or TiO_2 films have recently been employed [35].

HTM and ETM-free PSCs

To make device fabrication easier and lessen interface charge recombination, HTM- and ETM-free devices were constructed. It should be highlighted that using HTM-free inverted device architecture on both Pb- and Sn-based perovskites resulted in notable improvements in device performance and stability [36–38]. Using a Pb-Sn mixture, Yuqin et al. fabricated an inverted planar PSCs based on HTM-free $ITO/FAPb_{0.5}Sn_{0.5}I_3/PCBM/BCP/Al$ that produced a promising PCE of 7.94% [36]. On an ITO-coated glass substrate, Zhang et al. fabricated devices with and without PEDOT: PSS HTM and discovered that the HTM-free PSC is more stable than the devices with PEDOT: PSS [39]. With an inverted structure of

ITO/MAPbI3/C60/BCP/Ag, Li et al demonstrated hole conductor-free PSC. Devices had exceptional photovoltaic performance and stability, with reported PCE as high as 15.0% [37].

5.   High efficiency PSCs

The significant modification increased conversion efficiency to 17.9%-22.7% [4, 40]. Instead of using only MAPbI$_3$ as in the prior perovskite composition, a significant amount of a mixture of methylammonium lead bromide (MAPbBr$_3$) and formamidinium lead iodide (FAPbI$_3$) was used, as FAPbI$_3$ has a narrower bandgap (1.48 eV) and is closer to the Shockey-Queisser (S-Q) limit. It is possible to get a short-circuit current density (Jsc) of >26 mA cm^{-2} by fully utilising FAPbI$_3$'s wide absorption spectrum. Despite its benefits, FAPbI$_3$ alone produces a α-phase (a non-perovskite structure). δ-phase perovskite with high crystallinity was stabilised by adding 40 mol% methylammonium chloride (MACl) in order to address this issue [41].

The conversion efficiency increases to 23.3%–23.7% after advance modification [42]. The certification standard underwent a considerable revision during this time. Prior to this, the average values of the reverse scan and forward IV scan were used to certify results. The new measurement approach was known as the quasi-steady-state IV (QSS-IV), nevertheless. A new certification procedure was required where the PSC was subjected to ongoing light-soaking conditions since cell stability difficulties were still an issue. A QSS-IV sweep was carried out by fixing the bias voltage until the measured current stabilized (<0.07% change). Ten voltage points were employed, and the period between current observations was roughly 10 s. Measurements were typically performed for around 30 minutes under continuous light soaking. Due to the fact that TiO$_2$ ETL was known to exhibit undesirable photocatalytic effects when illuminated with UV light, PSCs with a TiO$_2$ ETL structure found it difficult to achieve a constant conversion efficiency. SnO$_2$ was used in place of TiO$_2$ in the ETL to resolve this issue. The conversion efficiency of a PSC with a SnO$_2$ ETL on an indium tin oxide (ITO) substrate was initially verified by you and your coworkers. They applied commercially available SnO$_2$ nanoparticles to the ITO substrate using the spin-coating technique. ITO's open-circuit voltage (Voc) and fill factor (FF) are higher, but Jsc is lower because ITO's sur surface is more level than FTO's. Even in the presence of constant light soaking, the SnO$_2$ ETL produced PSCs that were more stable.

Due to further developments, the conversion efficiency was raised to 24.2%–25.2% [43] and the cell shape and composition were significantly altered. FTO was employed structurally in place of ITO, and chemical bath deposition (CBD) SnO$_2$ was used in place of c-TiO_2/mp-TiO_2. The rough surface of FTO was utilised to scatter incoming radiation and trap light by increasing the optical path in order to decrease optical losses. To enhance the optoelectronic capabilities with a large grain size, $FAPbI_3$ and $MAPbBr_3$ (0.8 mol%) with MACl (ca. 35 mol%) and PbI_2 excess (9 mol%) are employed as the composition. When the $MAPbBr_3$ level reached at least 5 mol%, the $FAPbI_3$-based perovskite phase was regarded as stable. The stabilisation of the perovskite layer with improved grain size, carrier life time (>3.6 ms), and mobility (31.2 cm^2 V^{-1} s^{-1}) may be achieved, surprisingly, with only a tiny amount of $MAPbBr_3$ (1 mol%). With such a perfect active composition, the perovskite layer and ETL were able to control carriers very effectively. Even though there are severe troughs at the FTO surface, the SnO_2 layer at ETL forms complete coverage with interconnected domains thanks to CBD and careful pH and reaction time control. Using HTL that has been alkylammonium bromide-passivated, the hole carrier transfer has been enhanced. Passivation can thereby enhance carrier management by independently passivating the bulk with a tiny amount of $MAPbBr_3$ and the interface with 2D perovskite. It has been demonstrated that this carrier management strategy increases Voc and FF, decreases non-radiative recombination loss, and leads to >25% conversion efficiency.

How much PSC conversion efficiency can be improved in the future is of interest to researchers. In the near future, PSCs are anticipated to have conversion efficiencies greater than 26%. PSCs will trump the most efficient single-crystal silicon solar cells if their efficiency exceeds 26.1% [44]. PSCs have reached Jsc = 25.2 mA/cm^2, Voc = 1.18 V, and FF = 84.8%, according to the most current results reported in Nature. The measured Voc of 1.18 V suggests a voltage loss of 0.32 V, which is near to the theoretical value of 0.3 V if the bandgap of $(FAPbI_3)_{0.992}(MAPbBr_3)_{0.008}$ is 1.50 eV. There is, however, still opportunity for development. Since FF is 92% of the highest theoretical value, there is still room for development. In the instance of Jsc, 92% of the S-Q limit appears to be near to the highest value that may be achieved given that a glass substrate has an optical reflection loss of ~5%. However, optical advancements might boost Jsc much more. A better perovskite composition with the right number of additives would minimise the interface flaws to realise the best conversion efficiency using an ETL with uniform film coverage and thickness. It's interesting to note that while the HTL has not changed thus far, the ETL and perovskite layers have undergone continual compositional and structural changes. New HTLs that further improve the conversion efficiency and stability must be developed in order to realise commercial PSCs that maintain stability even under high temperature and humidity conditions.

6.   Perovskite powder production for diverse application

The preparation methods for perovskites are restricted for usage in other applications despite their extensive use in solar technology. The hybrid perovskite materials could not be used in as many different solar applications due to the limited preparation methods. The larger-scale synthesis of perovskite materials for a variety of uses still requires a simple and quick process. As a result, we have created a simple solution-based, quick, and large-scale powder production process. This novel technique takes less time and is more flexible for producing large quantities of organic-inorganic hybrid halide perovskite. Perovskites in powder form actually have significantly improved ambient stability, with exposure to ambient conditions (i.e., 30° C, 60% relative humidity) demonstrating barely any loss in their optical characteristics and crystal structures. It explains that the perovskite powder's high purity and crystallinity contributed to its improved ambient stability. The hybrid perovskites' powder form can be used for additional electronic applications, such as light-emitting diodes, lasers, photodetectors, and high-performance optoelectronic devices, in addition to their photovoltaic capabilities.

Due to their hygroscopic nature, perovskite precursors like MAI, MABr, FAI, and FABr are extremely moisture sensitive. Apart from the sensitivity of organic components, the hygroscopic character of perovskites (such $MAPbI_3$ and $FAPbI_3$) is the main cause of extrinsic instability, whereas the structure of the absorber material can be linked to intrinsic instability. We have devised a method for producing very stable perovskite powders on a wide scale for use over an extended period of time in order to increase stability. The new technique was used to create well-crystalline perovskite powders such as $MAPbI_3$, $MAPbI_3$, $FAPbI_3$, $FAPbBr_3$, $CsPbI_3$, and $CsPbBr_3$. A schematic representation of the production process for perovskite powder on a large scale is shown in Fig. 4(a). Only an equal amount of perovskite and chlorobenzene solution were used in the previous approach. However, a 1:10 vol ratio of perovskite solution and chlorobenzene was employed in this improved approach. The huge volume of the hot chlorobenzene solution is used for high purity with large-scale production in addition to quick crystallisation.

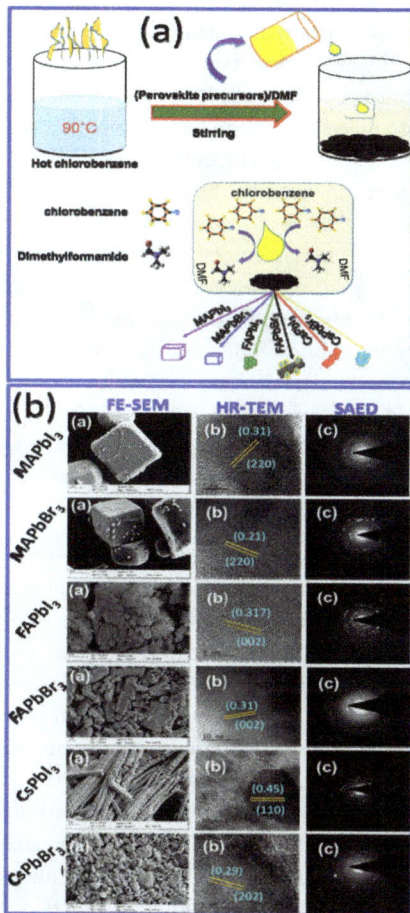

Figure 4. (a) Schematic representation of the production of perovskite powder in large-scale. The hot chlorobenzene and hot perovskite solution (1:10 vol ratio) were mixed together in the ambient air as shown in the schematic diagram. The entire perovskite production process was solution-based. (b) SAED pattern photos from FE-SEM and HR-TEM. For all FE-SEM photos, the scale-bar denotes a distance of 5 μm. Rapid crystallisation was used to prepare all of the perovskites using the solution process approach. The chlorobenzene was evaporated in a hot air oven at 100° C in an environment of air, and the perovskite granules were directly recovered from this process [45].

Fig. 4(b) shows FE-SEM and HR-TEM images corresponding SAED effectively retard the penetration of oxygen and moisture into the powders to prolong the lifetime of perovskite powders. The ionic radius differences between I- and Br-ions with 6-fold coordination, which are 2.2 and 1.96, respectively, appear to be the cause of the structural differences between iodine and bromide-based perovskites with the same organic cation or metal cation. Generally speaking, at room temperature, the photoactive black phase (α-FAPbI$_3$ and CsPbI$_3$) can phase transition readily into the stable yellow phase (α-phase). As a result, the phases of FAPbI$_3$ and CsPbI$_3$ were annealed for 10 minutes at 120 and 250 °C, respectively, for the XRD measurements. The remaining aged samples were used immediately for XRD analysis, and they are shown to have peaks that are very sharp and identical to those of the freshly created samples. It showed that the powdered perovskites have very excellent crystal stability. The yield for each powder is greater than 96%. The synthesis of other perovskites, such as compositional perovskites (MA$_x$FA$_{(1-x)}$PbI$_3$, Cs$_x$FA$_{(1-x)}$PbI$_3$, etc.), and other perovskites, such as EABX$_3$, GuBX$_3$, and RbBX$_3$, can also be done using this solution-based large-scale perovskite powder process.

7. Perovskite solar cells commercialization

The National Renewable Energy Laboratory (NREL)-certified Oxford Photovoltaics (Oxford PV) perovskite-on-silicon tandem solar cell achieves a world-record efficiency of 29.5% in 2022 [46]. This not only surpasses the previous record set by the corporation of 28.0%, but also the Helmholtz Zentrum Berlin (HZB) performance record of 29.2%. The perovskite-on-silicon tandem arrangement offers a more accessible method to develop PSCs in the current photovoltaic industry, which has been viewed as extremely promising in the industrial process of PSCs, as opposed to replacing silicon solar cells. One of the earliest solar businesses, Oxford PV was founded in 2010 from the University of Oxford and focuses on the commercialization of perovskite-based photovoltaic technology. The popular perovskite-on-silicon tandem solar cell technique enables for performance gains above what was previously possible for cell modules. With an anticipated combined cell efficiency of over 30% and its simplicity of integration, the perovskite-based tandem technology has demonstrated significant potential to be compatible with the present solar sector.

At the same time, Chinese solar enterprises have focused on increasing PSC manufacturing. To break the previous world record set by Toshiba, Hangzhou Microquanta Co. Ltd. initially developed the production line of 20 MW using 200-800 cm perovskite solar cell modules and achieved 11.98% PCE. 2019 saw the completion of the material synthesis and manufacturing process for large-area perovskite modules by Suzhou GCL Nano Technology Co., Ltd. (GCL Nano), a division of GCL group. Later, GCL stated plans to create a product line for a 100MW production line and to carry out perovskite module commercial production. GCL Nano stated that it would achieve a PCE of 15.21% on a 1,241.16 cm^2 effective area. In addition, the Jayu Group and its subsidiary Jayu Solar Energy Technology Co., Ltd. give the commercialization of third-generation perovskite photovoltaic technology a lot of attention and are planning a new architecture for their future manufacturing facilities. In order to reach carbon neutrality in 2060, it is planned that non-fossil energy should make up 25% of all energy used in 2030 and that the capacity of renewable energy should expand to 1.2 billion KW or more. Compared to 50 million KW annually over the previous five years, the photovoltaic industry should reach 70-90 million KW annually by the end of the next decade.

Since first reports, perovskite solar cells (PSCs) have significantly improved in stability and efficiency; but, as will be covered in this road map, there are still issues that need to be addressed before they can be commercialised. However, a number of businesses are engaged in PSC research, including Oxford Photovoltaic (PV), Utmo Light Ltd, Imec, Microquanta Semiconductor, Solliance, Toshiba, Saule Technologies, Wonder Solar Ltd, GCL New Energy, Xeger Sweden AB, Alta Devices, G24 Power Ltd, FlexLink Systems, Polyera Corporation Solar Print Ltd, New Energy Technologies Inc. Korver Corp., Solar-Tectic, Ubiquitous Energy Inc [47, 48]. Some of these businesses, like Oxford PV (125 MW capacity production line in partnership with Meyer Burger) and Saule Technologies, have been setting up new pilot production lines and/or increasing their production capacity [47, 48], and commercial prototypes have also been installed (72 modules in Saule Technologies' innovative Japanese hotel, Henn-Na [47]). Additionally, GCL New Energy is constructing a 100 MW production line in Kunshan while Microquanta Semiconductor has already constructed a 5 GW production line [48]. Printing technologies (Saule Technologies and Wonder Solar Ltd) are also being commercialised [47, 48], and Wonder Solar Ltd has already shown off a 110 m^2 outdoor power generation system. Additionally, a number of collaborations and consortiums comprising national laboratories, businesses, research centres, and universities have been formed, including US-MAP (Manufacturing of Advanced Perovskites) [50] and the European Perovskite Initiative (EPKI) [49].

8.   Future perspective and challenges of the perovskite

In addition to other factors like high efficiency, simple manufacture, and low cost, stability is a key component of the commercialization of PSCs. However, because different experiments use different testing settings, such as humidity, temperature, and encapsulation technique, the stability results offered by diverse researchers cannot be accurately compared. In contrast to stability-related features like lifetime and deterioration rates, PCE is a well-defined metric that can be validated in accordance with standards. It is also imperative to standardise the necessary test conditions for PSC stability testing, with a focus on mechanical stability, thermal stability, device hysteresis, and stability under exposure to light, moisture, and oxygen for each fabrication procedure. It should be mentioned that while inorganic solar cell stability tests (ISOS) have methodologies that have been authorised, they are rarely used to examine the stability of PSCs.

Additionally, R&D goals should be established for PSC development and successful commercialization in order to meet market demand. A useful method to comprehend the ranking of PSC traits for successful commercialization is the Keno model, which is imposed in Fig. 5 [51]. The Kano model stands out for its tight emphasis on consumer perception; existing and/or potential new product features are graded according to the potential level of customer satisfaction they may bring. Such categorised PSCs qualities are shown in the figure as being required for commercialisation in the current solar cell market. Strong research initiatives in these areas are necessary for market penetration. In contrast to "must be's," which are the fundamental needs for the technology as well as the absolute necessities for market access, "one-dimensional" specifications have a linear relationship to the quality and value of the product. Due to the considerable decrease in silicon solar cell manufacturing prices over the past few years, photovoltaics can now be used as a large-scale sustainable energy source to displace fossil fuels. Numerous studies indicate that, in well-equipped places with enough sun light, solar energy may completely compete with conventional plants when the cost is less than 32 cents/W. However, due to thermalization loss, low absorption coefficient, and high material needs, silicon-based photovoltaic technology has reached the point where it is no longer cost-effective to use it. PSCs offers chances to develop solar technology that goes well beyond silicon-based systems. PSCs can be produced using raw materials that are relatively modest in purity (> 99%), in contrast to silicon-based photovoltaic technology, which needs extremely pure raw materials (99.999%).

More importantly, the low temperature solution methods used to create perovskite films lead to reduced power assumption and environmental pollution. Despite having greater PCEs of roughly 30%, the second-generation gallium arsenide thin-film solar cell is very expensive. The cost of producing perovskite modules is anticipated to be 50% less than that of monocrystalline silicon. However, due to the lack of a proven production line, silicon solar cells are hard to replace in the photovoltaic market. As an alternative, one of the technological paths leading to the industrial advancement of PSCs should be the perovskite-on-silicon tandem structure. Although there are some difficulties with transparent conductive layers, tunnel junctions, and light utilisation, perovskites are good wide-bandgap materials in silicon-based multijunction solar cells because of how easily their bandgap can be tuned and because they are compatible with compatible techniques

[52]. The manufacturing supply chain is shorter when PSCs are commercialised, and large-scale coating equipment can be produced quickly. In around 3 years, the key technologies for flexible large-area perovskite solar photovoltaic cells will be developed. With multiple efforts in the academic and industrial fields, the pathways of PSCs from the laboratory to the market are beginning to emerge. According to estimates, the photovoltaic market will grow by at least $100 billion as a result of the commercialization of PSCs.

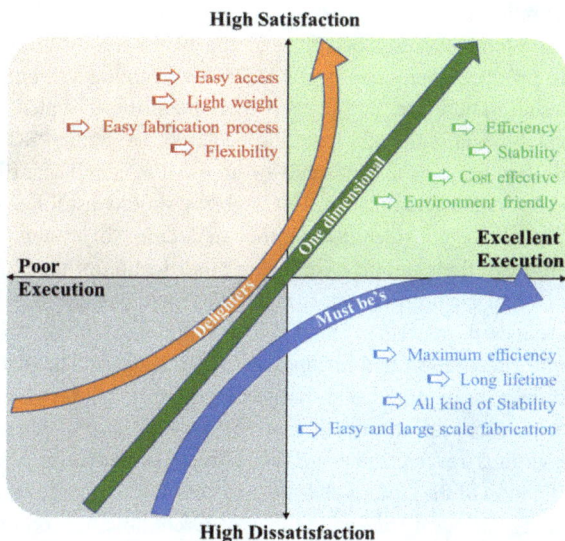

Figure 5 Keno model for proposal of successfully commercialization of PSCs [55].

Conclusions

Over the past ten years, the solar industry has paid close attention to PSCs due to their rapid development. The device efficiency of PSCs has surpassed polycrystalline Si cells and is now nearing that of single crystal Si cells, reaching a world-record value of 25.5%. However, conventional Si technology, which has been proven to be reliable for more than 25 years, continues to rule the solar business. One of the primary barriers to PSCs entering the photovoltaic sector is their lack of operational reliability. Normally, both extrinsic and intrinsic factors have a significant impact on the device stability of PSCs. By using sophisticated encapsulation procedures, extrinsic environmental variables like moisture and oxygen in the surrounding air can be totally eliminated. As a result, we provide a thorough analysis of the elements—such as ion dissociation and migration, metal-perovskite interactions, and residual strain-induced perovskite degradation—that affect the intrinsic stability of perovskite materials in this study.

The superb and tuneable optoelectronic features of materials, which have been disclosed through decades of research in the PV sector, have been a driving force behind the recent rapid development of PSCs. Recent research indicates that the main concerns for PSCs are no longer efficiency or cost; however, the short operational lifetime under typical environmental conditions is a major barrier to commercialization. It is particularly difficult to obtain the long-term stability of PSCs in the commercialization of perovskite PV technology or to produce operationally stable PSCs due to the soft and ionic character of perovskite materials. Therefore, before their manufacturing is scaled up, the stability of PSCs must be established in the lab. It should be noted that while PSC stability has lately increased from a few minutes to thousands of hours, this is still insufficient for practical implementation. For commercialization, PSC stability should last for at least ten years. The various researchers have considered a number of factors for systematic engineering of perovskites to improve their stability, including modification of the structural design, use of various charge and electron transport materials, including various metal-oxide films, different hole transport materials that have hydrophobic nature, modification in the electrode material preparation and encapsulation procedures. The stability of PSCs has lately increased from a few minutes to thousands of hours, but this is still insufficient for practical implementation; for commercialization, the stability of PSCs need stretch beyond ten years. The various researchers have taken into consideration a number of factors, including modification of the structural design, use of various charge and electron transport materials, including various metal-oxide films, different hole transport materials that have hydrophobic nature, modification in the electrode material preparation and encapsulation procedures, for systematic engineering of perovskites to improve their stability.

Reference

[1]   A. Kojima, K. Teshima, Y. Shirai, T. Miyasaka, Organometal Halide Perovskites as Visible-Light Sensitizers for Photovoltaic Cells, Journal of the American Chemical Society. 131 (2009) 6050–6051. https://doi.org/10.1021/ja809598r

[2]   M.M. Lee, J. Teuscher, T. Miyasaka, T.N. Murakami, H.J. Snaith, Efficient Hybrid Solar Cells Based on Meso-Superstructured Organometal Halide Perovskites, Science. 338 (2012) 643–647. https://doi.org/10.1126/science.1228604

[3]   J. Burschka, N. Pellet, S.-J. Moon, R. Humphry-Baker, P. Gao, M.K. Nazeeruddin, M. Grätzel, Sequential deposition as a route to high-performance perovskite-sensitized solar cells, Nature. 499 (2013) 316–319. https://doi.org/10.1038/nature12340

Materials Research Forum LLC
https://doi.org/10.21741/9781644903032-6

[4]   N.J. Jeon, J.H. Noh, W.S. Yang, Y.C. Kim, S. Ryu, J. Seo, S. Il Seok, Compositional engineering of perovskite materials for high-performance solar cells, Nature. 517 (2015) 476–480. https://doi.org/10.1038/nature14133

[5]   P. Kajal, K. Ghosh, S. Powar, Manufacturing Techniques of Perovskite Solar Cells, in: 2018: pp. 341–364. https://doi.org/10.1007/978-981-10-7206-2_16

[6]   A.M. Elseman, A.H. Zaki, A.E. Shalan, M.M. Rashad, Q.L. Song, TiO2 Nanotubes: An Advanced Electron Transport Material for Enhancing the Efficiency and Stability of Perovskite Solar Cells, Industrial & Engineering Chemistry Research. 59 (2020) 18549–18557. https://doi.org/10.1021/acs.iecr.0c03415

[7]   N. kour, R. Mehra, Chandni, Efficient design of perovskite solar cell using mixed halide and copper oxide, Chinese Physics B. 27 (2018) 18801. https://doi.org/10.1088/1674-1056/27/1/018801

[8]   S. Bansal, P. Aryal, Evaluation of new materials for electron and hole transport layers in perovskite-based solar cells through SCAPS-1D simulations, in: 2016 IEEE 43rd Photovoltaic Specialists Conference (PVSC), 2016: pp. 747–750. https://doi.org/10.1109/PVSC.2016.7749702

[9]   C.Y. Xu, W. Hu, G. Wang, L. Niu, A.M. Elseman, L. Liao, Y. Yao, G. Xu, L. Luo, D. Liu, G. Zhou, P. Li, Q. Song, Coordinated Optical Matching of a Texture Interface Made from Demixing Blended Polymers for High-Performance Inverted Perovskite Solar Cells, ACS Nano. 14 (2020) 196–203. https://doi.org/10.1021/acsnano.9b07594

[10]   M. Yang, D.H. Kim, T. Klein, Z. Li, M. Reese, B. Tremolet de Villers, J. Berry, M.F.A.M. Hest, K. Zhu, Highly Efficient Perovskite Solar Modules by Scalable Fabrication and Interconnection Optimization, ACS Energy Letters. 3 (2018). https://doi.org/10.1021/acsenergylett.7b01221

[11]   A.R. Pascoe, S. Meyer, W. Huang, W. Li, I. Benesperi, N.W. Duffy, L. Spiccia, U. Bach, Y.-B. Cheng, Enhancing the Optoelectronic Performance of Perovskite Solar Cells via a Textured CH3NH3PbI3 Morphology, Advanced Functional Materials. 26 (2016) 1278–1285. https://doi.org/https://doi.org/10.1002/adfm.201504190

[12]   V.M. Goldschmidt, Die Gesetze der Krystallochemie, Naturwissenschaften. 14 (1926) 477–485. https://doi.org/10.1007/BF01507527

[13]   G. Kieslich, S. Sun, A.K. Cheetham, Solid-state principles applied to organic–inorganic perovskites: new tricks for an old dog, Chemical Science. 5 (2014) 4712–4715. https://doi.org/10.1039/C4SC02211D

[14]   C.C. Stoumpos, M.G. Kanatzidis, The Renaissance of Halide Perovskites and Their Evolution as Emerging Semiconductors, Accounts of Chemical Research. 48 (2015) 2791–2802. https://doi.org/10.1021/acs.accounts.5b00229

[15]   C. Li, X. Lu, W. Ding, L. Feng, Y. Gao, Z. Guo, Formability of ABX3 (X = F, Cl, Br, I) halide perovskites, Acta Crystallographica Section B. 64 (2008) 702–707. https://doi.org/https://doi.org/10.1107/S0108768108032734

[16]   N. McKinnon, D. Reeves, M. Akabas, 5-HT3 receptor ion size selectivity is a property of the transmembrane channel, not the cytoplasmic vestibule portals, The Journal of General Physiology. 138 (2011) 453–466. https://doi.org/10.1085/jgp.201110686

[17]   G.E. Eperon, S.D. Stranks, C. Menelaou, M.B. Johnston, L.M. Herz, H.J. Snaith, Formamidinium lead trihalide: a broadly tunable perovskite for efficient planar heterojunction solar cells, Energy & Environmental Science. 7 (2014) 982–988. https://doi.org/10.1039/C3EE43822H

[18]   P.-P. Sun, Q.-S. Li, L.-N. Yang, Z.-S. Li, Theoretical insights into a potential lead-free hybrid perovskite: substituting Pb2+ with Ge2+, Nanoscale. 8 (2016) 1503–1512. https://doi.org/10.1039/C5NR05337D

[19]   A.A. Bakulin, O. Selig, H.J. Bakker, Y.L.A. Rezus, C. Müller, T. Glaser, R. Lovrincic, Z. Sun, Z. Chen, A. Walsh, J.M. Frost, T.L.C. Jansen, Real-Time Observation of Organic Cation Reorientation in Methylammonium Lead Iodide Perovskites, The Journal of Physical Chemistry Letters. 6 (2015) 3663–3669. https://doi.org/10.1021/acs.jpclett.5b01555

[20]   J.M. Frost, K.T. Butler, F. Brivio, C.H. Hendon, M. van Schilfgaarde, A. Walsh, Atomistic Origins of High-Performance in Hybrid Halide Perovskite Solar Cells, Nano Letters. 14 (2014) 2584–2590. https://doi.org/10.1021/nl500390f

[21]   J.-W. Lee, D.-J. Seol, A.-N. Cho, N.-G. Park, High-Efficiency Perovskite Solar Cells Based on the Black Polymorph of HC(NH2)2PbI3, Advanced Materials. 26 (2014) 4991–4998. https://doi.org/https://doi.org/10.1002/adma.201401137

[22]   G.E. Eperon, G.M. Paternò, R.J. Sutton, A. Zampetti, A.A. Haghighirad, F. Cacialli, H.J. Snaith, Inorganic caesium lead iodide perovskite solar cells, Journal of Materials Chemistry A. 3 (2015) 19688–19695. https://doi.org/10.1039/C5TA06398A

[23]   M. Kulbak, D. Cahen, G. Hodes, How Important Is the Organic Part of Lead Halide Perovskite Photovoltaic Cells? Efficient CsPbBr 3 Cells, The Journal of Physical Chemistry Letters. 6 (2015) 150610174239009. https://doi.org/10.1021/acs.jpclett.5b00968

[24]   M.R. Filip, G.E. Eperon, H.J. Snaith, F. Giustino, Steric engineering of metal-halide perovskites with tunable optical band gaps, Nature Communications. 5 (2014) 5757. https://doi.org/10.1038/ncomms6757

[25]   J.H. Noh, S.H. Im, J.H. Heo, T.N. Mandal, S. Il Seok, Chemical Management for Colorful, Efficient, and Stable Inorganic–Organic Hybrid Nanostructured Solar Cells, Nano Letters. 13 (2013) 1764–1769. https://doi.org/10.1021/nl400349b

[26]   M. Saliba, T. Matsui, J.-Y. Seo, K. Domanski, J.-P. Correa-Baena, M.K. Nazeeruddin, S.M. Zakeeruddin, W. Tress, A. Abate, A. Hagfeldt, M. Grätzel, Cesium-containing triple cation perovskite solar cells: improved stability, reproducibility and high efficiency, Energy & Environmental Science. 9 (2016) 1989–1997. https://doi.org/10.1039/C5EE03874J

[27]   J. Jeong, M. Kim, J. Seo, H. Lu, P. Ahlawat, A. Mishra, Y. Yang, M.A. Hope, F.T. Eickemeyer, M. Kim, Y.J. Yoon, I.W. Choi, B.P. Darwich, S.J. Choi, Y. Jo, J.H. Lee, B. Walker, S.M. Zakeeruddin, L. Emsley, U. Rothlisberger, A. Hagfeldt, D.S. Kim, M. Grätzel, J.Y. Kim, Pseudo-halide anion engineering for α-FAPbI3 perovskite solar cells, Nature. 592 (2021) 381–385. https://doi.org/10.1038/s41586-021-03406-5

[28]   T. Salim, S. Sun, Y. Abe, A. Krishna, A.C. Grimsdale, Y.M. Lam, Perovskite-based solar cells: impact of morphology and device architecture on device performance, Journal of Materials Chemistry A. 3 (2015) 8943–8969. https://doi.org/10.1039/C4TA05226A

[29]   M.I. Asghar, J. Zhang, H. Wang, P.D. Lund, Device stability of perovskite solar cells – A review, Renewable and Sustainable Energy Reviews. 77 (2017) 131–146. https://doi.org/https://doi.org/10.1016/j.rser.2017.04.003

[30]   G.-H. Kim, D.S. Kim, Development of perovskite solar cells with >25% conversion efficiency, Joule. 5 (2021) 1033–1035. https://doi.org/https://doi.org/10.1016/j.joule.2021.04.008

[31]   D. Yang, R. Yang, J. Zhang, Z. Yang, S. (Frank) Liu, C. Li, High efficiency flexible perovskite solar cells using superior low temperature TiO2, Energy & Environmental Science. 8 (2015) 3208–3214. https://doi.org/10.1039/C5EE02155C

[32]   T. Ibn-Mohammed, S.C.L. Koh, I.M. Reaney, A. Acquaye, G. Schileo, K.B. Mustapha, R. Greenough, Perovskite solar cells: An integrated hybrid lifecycle assessment and review in comparison with other photovoltaic technologies, Renewable and Sustainable Energy Reviews. 80 (2017) 1321–1344. https://doi.org/https://doi.org/10.1016/j.rser.2017.05.095

[33]   J.-Y. Jeng, Y.-F. Chiang, M.-H. Lee, S.-R. Peng, T.-F. Guo, P. Chen, T.-C. Wen, CH3NH3PbI3 Perovskite/Fullerene Planar-Heterojunction Hybrid Solar Cells, Advanced Materials. 25 (2013) 3727–3732. https://doi.org/https://doi.org/10.1002/adma.201301327

[34]   Q. Dong, Y. Yuan, Y. Shao, Y. Fang, Q. Wang, J. Huang, Abnormal crystal growth in CH3NH3PbI3−xClx using a multi-cycle solution coating process, Energy & Environmental Science. 8 (2015) 2464–2470. https://doi.org/10.1039/C5EE01179E

[35]   W. Chen, Y. Wu, Y. Yue, J. Liu, W. Zhang, X. Yang, H. Chen, E. Bi, I. Ashraful, M. Grätzel, L. Han, Efficient and stable large-area perovskite solar cells with inorganic charge extraction layers, Science. 350 (2015) 944–948. https://doi.org/10.1126/science.aad1015

[36]   Y. Liao, X. Jiang, W. Zhou, Z. Shi, B. Li, Q. Mi, Z. Ning, Hole-transporting layer-free inverted planar mixed lead-tin perovskite-based solar cells, Frontiers of Optoelectronics. 10 (2017) 103–110. https://doi.org/10.1007/s12200-017-0716-6

[37]   Y. Li, S. Ye, W. Sun, W. Yan, Y. Li, Z. Bian, Z. Liu, S. Wang, C. Huang, Hole-conductor-free planar perovskite solar cells with 16.0% efficiency, Journal of Materials Chemistry A. 3 (2015) 18389–18394. https://doi.org/10.1039/C5TA05989E

[38]   W. Ke, G. Fang, J. Wan, H. Tao, Q. Liu, L. Xiong, P. Qin, J. Wang, H. Lei, G. Yang, M. Qin, X. Zhao, Y. Yan, Efficient hole-blocking layer-free planar halide perovskite thin-film solar cells, Nature Communications. 6 (2015) 6700. https://doi.org/10.1038/ncomms7700

[39]   Y. Zhang, X. Hu, L. Chen, Z. Huang, Q. Fu, Y. Liu, L. Zhang, Y. Chen, Flexible, hole transporting layer-free and stable CH3NH3PbI3/PC61BM planar heterojunction perovskite solar cells, Organic Electronics. 30 (2016) 281–288. https://doi.org/https://doi.org/10.1016/j.orgel.2016.01.002

[40]   E.H. Jung, N.J. Jeon, E.Y. Park, C.S. Moon, T.J. Shin, T.-Y. Yang, J.H. Noh, J. Seo, Efficient, stable and scalable perovskite solar cells using poly(3-hexylthiophene), Nature. 567 (2019) 511–515. https://doi.org/10.1038/s41586-019-1036-3

[41]   M. Kim, G.-H. Kim, T.K. Lee, I.W. Choi, H.W. Choi, Y. Jo, Y.J. Yoon, J.W. Kim, J. Lee, D. Huh, H. Lee, S.K. Kwak, J.Y. Kim, D.S. Kim, Methylammonium Chloride Induces Intermediate Phase Stabilization for Efficient Perovskite Solar Cells, Joule. 3 (2019) 2179–2192. https://doi.org/10.1016/j.joule.2019.06.014

[42]   Q. Jiang, Y. Zhao, X. Zhang, X. Yang, Y. Chen, Z. Chu, Q. Ye, X. Li, Z. Yin, J. You, Surface passivation of perovskite film for efficient solar cells, Nature Photonics. 13 (2019) 460–466. https://doi.org/10.1038/s41566-019-0398-2

Materials Research Forum LLC
https://doi.org/10.21741/9781644903032-6

[43]   J.J. Yoo, G. Seo, M.R. Chua, T.G. Park, Y. Lu, F. Rotermund, Y.-K. Kim, C.S. Moon, N.J. Jeon, J.-P. Correa-Baena, V. Bulović, S.S. Shin, M.G. Bawendi, J. Seo, Efficient perovskite solar cells via improved carrier management, Nature. 590 (2021) 587–593. https://doi.org/10.1038/s41586-021-03285-w

[44]   NREL (2021). Best Research-Cell Efficiency Chart. https://www.nrel.gov/pv/cell-efficiency.

[45]   G. Murugadoss, M. Rajesh Kumar, V.M. Shanmugam, Rational design and development of perovskite materials: Analysis of structural, optical, morphological and phase transition, Materials Science in Semiconductor Processing. 117 (2020) 105177. https://doi.org/https://doi.org/10.1016/j.mssp.2020.105177

[46]   A. Al-Ashouri, E. Köhnen, B. Li, A. Magomedov, H. Hempel, P. Caprioglio, J.A. Márquez, A.B. Morales Vilches, E. Kasparavicius, J.A. Smith, N. Phung, D. Menzel, M. Grischek, L. Kegelmann, D. Skroblin, C. Gollwitzer, T. Malinauskas, M. Jošt, G. Matič, B. Rech, R. Schlatmann, M. Topič, L. Korte, A. Abate, B. Stannowski, D. Neher, M. Stolterfoht, T. Unold, V. Getautis, S. Albrecht, Monolithic perovskite/silicon tandem solar cell with >29% efficiency by enhanced hole extraction, Science. 370 (2020) 1300–1309. https://doi.org/10.1126/science.abd4016

[47]   P. Roy, N. Kumar Sinha, S. Tiwari, A. Khare, A review on perovskite solar cells: Evolution of architecture, fabrication techniques, commercialization issues and status, Solar Energy. 198 (2020) 665–688. https://doi.org/https://doi.org/10.1016/j.solener.2020.01.080

[48]   T. Wu, Z. Qin, Y. Wang, Y. Wu, W. Chen, S. Zhang, M. Cai, S. Dai, J. Zhang, J. Liu, Z. Zhou, X. Liu, H. Segawa, H. Tan, Q. Tang, J. Fang, Y. Li, L. Ding, Z. Ning, Y. Qi, Y. Zhang, L. Han, The Main Progress of Perovskite Solar Cells in 2020–2021, Nano-Micro Letters. 13 (2021) 152. https://doi.org/10.1007/s40820-021-00672-w

[49]   European Perovskite Initiative (available at: https://epki.eu/).

[50]   U.S. MAP -Manufacturing of Advanced Perovskites (available at: www.usa-perovskites.org/index.html)

[51]   T.A. Chowdhury, M.A. Bin Zafar, M. Sajjad-Ul Islam, M. Shahinuzzaman, M.A. Islam, M.U. Khandaker, Stability of perovskite solar cells: issues and prospects, RSC Advances. 13 (2023) 1787–1810. https://doi.org/10.1039/D2RA05903G

[52]   H. Li, W. Zhang, Perovskite Tandem Solar Cells: From Fundamentals to Commercial Deployment, Chemical Reviews. 120 (2020) 9835–9950. https://doi.org/10.1021/acs.chemrev.9b00780

Keyword Index

b

www.ingramcontent.com/pod-product-compliance
Lightning Source LLC
Chambersburg PA
CBHW071235210326
41597CB00016B/2063